POPULAR

ERRORS

POPULAR ERRORS

LAURENT JOUBERT

TRANSLATED AND ANNOTATED BY
GREGORY DAVID de ROCHER

THE UNIVERSITY OF ALABAMA PRESS

•

TUSCALOOSA

Manufactured in the United States of America

Library of Congress Cataloging-in-Publication Data

Joubert, Laurent, 1529–1583.
Popular errors.

Translation of: Erreurs populaires au fait de la médecine et régime de santé, lre partie.
Bibliography: p.
Includes index.
1. Medical errors—Early works to 1800. I. Title.
R729.8.J6813 1989 610 88–1323
ISBN 0-8173-5289-9

British Library
Cataloguing-in-Publication
Data available

"Plus sapit vulgus quia tantum,
quantum opus est, sapit."

—LACTANTIUS

•

"This is what is related concerning such things.
Let us now see what is to be believed."

—LAURENT JOUBERT

•

"Primus sapientiae gradus est
falsa intelligere."

—LACTANTIUS

CONTENTS

ACKNOWLEDGMENTS

I wish to express my gratitude to the Research Grants Committee of The University of Alabama for grants-in-aid that have made it possible for me to gather materials for this translation. I am also indebted to Richard E. Peck, Dean of the College of Arts and Sciences, and to Roger E. Sayers, Academic Vice President, for a research leave allowing the preparation of the manuscript. I wish to extend special thanks to Richard W. Baldes, Professor of Classical Languages, and to Alexandre and Marguerite Todorov, Doctors of Medicine, without whose aid annotating a Renaissance medical treatise would have proved too formidable. For the painstaking work of reading the manuscript I am indebted to Professor Barbara C. Bowen of the University of Illinois and to Professor Leona B. LeBlanc of Florida State University. Professor John P. Hermann of The University of Alabama helped wrestle my Gallicisms into more readable English, and Trinket Shaw's careful editing caught several errors and countless infelicities. The flaws that remain are attributable only to my own oversights or stubbornness.

TRANSLATOR'S PREFACE

The translation of the title of Joubert's *Erreurs populaires* might well have been *Popular Misconceptions.* At least, that was the one I had chosen before happening upon Natalie Zemon Davis's impressive study, *Society and Culture in Early Modern France,* which devotes considerable space to Joubert. Because she set a precedent among numerous readers who know her more direct, more word-for-word rendering of the title of Joubert's work, I have adopted it.

It is a translator's fantasy to be able to capture every nuance, every flavor of the time. This is, unhappily, impossible. Translation is at every moment compromise. I have nevertheless sought to render Joubert's sense faithfully (a word the author uses frequently), while espousing as much as possible his turns of phrase. I would have preferred an opposing page translation, as I would have for the *Treatise on Laughter*; unfortunately, the size of these works precludes such presentation.

The system of notes has been greatly simplified: all notes, both Joubert's and my own, appear at the end of the translation. Marginal notes by Joubert are in each case designated as such, with a translation of the note following the colon (e.g., [25]Joubert's note: Pliny, *Natural History*, Book 14, chap. 12.). In some cases my own comments follow, set off in a new paragraph.

G. de R.

INTRODUCTION

LAURENT JOUBERT AND THE *ERREURS POPULAIRES*

Laurent Joubert is no longer the relatively unknown figure he was fifteen years ago. Although a brief biblio-biographical study by P.-J. Amoreux was published in 1814, another by J. L. V. Broussonnet in 1829, and a long article on the *Erreurs populaires* by E. Wickersheimer in 1906, Joubert's name and work have never been well known, not even to specialists.[1] It is only fairly recently that books by Natalie Zemon Davis and Robert Cottrell, along with articles by Louis Dulieu, Claude Longeon, and Jean-Louis Gourg, have gained a place for Joubert in standard bibliographies.[2]

Yet, although he is beginning to interest students of intellectual history, literature, medical history, psychology, and law, the greater portion of his work is inaccessible. Joubert's *Traité du ris* (1579) is available in facsimile reprint, but until the publication of the annotated translation in 1980 (*Treatise on Laughter*), it could be read only by specialists. The *Erreurs populaires* is undoubtedly his most important book. I seek to supply both an annotated translation of this work and an account of the variants forced upon Joubert by his many critics. But other works by Joubert that are bound to be of great interest have yet to come to light: among them there is a treatise on gunshot wounds, a treatise on the plague, a pharmacopoeia, two volumes of medical and philosophical paradoxes, and a treatise on syphilis that, in all likelihood, came from his pen. In short, there is in him a wealth of Renaissance knowledge, practice, and mythology yet to be tapped.

The following brief biographical sketch is gleaned from P.-J. Amoreux and Louis Dulieu. Joubert was born on December 16, 1529, in the old province of Dauphiné in south central France. Although information on his early years is sparse, he was probably educated in his native city. At the age of twenty-one he went to Montpellier, a city renowned for the study of medicine. The student of Guillaume Rondelet, chancellor of the Faculté de Médecine, Joubert received his doctorate in 1558. He had acquired a reputation as a great pedagogue. We read, for example, that after Rondelet's death in 1556—only eight years after the beginning of his career—he was made regent as a result of a petition signed by the Montpellier students. A short time later he was appointed chancellor of the Faculté. In this capacity he attracted the attention of Catherine de' Medici, who called him to be her personal physician. The high point in his professional life came when he was made one of the king's physicians, *médecin ordinaire du roi.* He probably died in the small village of Lombers on October 21, 1582, not far from Montpellier, the city that had witnessed his rise to fame. Legend has it that, refusing to shirk his duty, he went out on a stormy night to treat a patient and never returned.[3]

Busy as he was with the obligations of his practice, Joubert found time to compose several medical works. He wrote voluminously in Latin and in French, edited the important *Grande chirurgie* of Guy de Chauliac, and published his *Erreurs populaires* to considerable scandal. In a work dedicated to Princess Marguerite de France he spoke openly of sexual matters and, what was worse, revealed in the vulgar tongue medical secrets previously shrouded in Latin. Three years before his death Joubert published his first written and most cherished work, the *Traité du ris*, or *Treatise on Laughter*. A considerable portion of Joubert's work appeared in print just before 1580: the *Grande chirurgie*, the *Erreurs populaires*, the *Traitté des arcbusades*, the *Question vulgaire* in 1578, and the *Traité du Ris* and the *Pharmacopea* in 1579.

Besides these works Joubert wrote the *Medicinae practicae priores* (in three books), the *Isagoge therapeutices methodi*, and the *De affectibus internis partium thoracis, tractatus alter*, all of which appeared in a single volume published in Lyons in 1577 and later in his *Opera latina*, which appeared in Lyons in 1582. An earlier edition of this work, published by J. Gregorii, is held in the Bibliothèque

Municipale de Bordeaux. Joubert's *Opuscula*, brief works published by his students, appeared in Lyons in 1571 as well as in his *Opera latina*; they contain his *Annotationes in duos priores libros Galeni, de facultatibus naturalibus*; *Annotationes in Galeni librum, de differentiis morborum*; *Annotationes in Galeni librum, de differentiis symptomarum*; *De convulsionis essentia et causis*; *De cerebri affectibus*; *Ars componendi medicamenta*; *De syruporum conficiendi modo, et retendi ratione*; *Quaestiones medicae pro regia disputatae*; and *Joannis Hucheri pro philosophia libertate oratio*. The second part of Joubert's *Opera latina* contains, with the exception of his tract on urine (*De urinarum differentiis, causis et judiciis*), letters and eulogies in which various medical questions are discussed, as well as health practices from antiquity. The last of Joubert's separate works to be published was his *Traité des eaux*, which appeared in Paris in 1603, more than twenty years after his death.

Yet these works, even the *Treatise on Laughter*, played no part in the immediate and resounding controversy over the publication of the *Erreurs populaires*. The reaction to the book can be divided into two distinct parts, one responding to Joubert's dedication of the *Erreurs populaires*, and the other to perceived threats to those with vested interests in medicine yet who were not physicians. The longest chapter of the treatise, chapter IV of Book Five, treats the topic "Whether There Is Certain Knowledge of the Virginity of a Maiden." This chapter made for much of the scandal as well as the success of the work. It was no sooner published in 1578 than its dedicatory letter to the young Princess Marguerite de France was seized upon as a notorious example of poor judgment as well as an uncommon breach of propriety. This criticism was launched mainly by Scévole de Sainte-Marthe. Even one of Joubert's staunch supporters, François Grudé de La Croix Du Maine (Joubert's contemporary and one of our most valuable sources of sixteenth-century bibliographical data), regrets the author's *manière libre*. The more general reaction to the treatise's publication came from the lower ranks of the medical profession. Midwives, apothecaries, and barber-surgeons rose to protect their interests, since Joubert had attacked them openly in his attempt to underscore the power of the physician in Renaissance France. Empirics such as the surgeon Ambroise Paré and Louyse Bourgeois were starting to enjoy considerable prestige, while the long-standing institutions of midwifery and apothecary practice came to be viewed by the Renaissance doctors as a menace to their privileged status.

The *Erreurs populaires*, then, can be read not only as an error book but as a frantic attempt at parrying the fierce blows to the power and prestige of physicians dealt by practical medicine. Respect for the authority of the ancients and for the role of physicians as their sole interpreters was starting to wane. Recipe and remedy books, as Natalie Zemon Davis points out, could not be published rapidly enough to satisfy a clientele that, seeking to become more self-sufficient, preferred books to visitations and prescriptions. Indeed, several of Joubert's works seek to confirm the institution of the physicians through open or veiled attacks against barber-surgeons (*Traitté des arcbusades*), apothecaries (*Pharmacopea* and *Erreurs populaires*, Book One), and midwives (*Erreurs populaires*, Books Two, Three, Four, and Five).

Still, to explain away the *Erreurs populaires* through a narrowly economic reading would be an impoverishment. The evidence does not show that Joubert is only attempting to raise the status of physicians by debunking common adages. On the contrary, Joubert relentlessly pursues hidden meanings, which he called truths. He rarely mocks popular sayings, even the most patently absurd; rather, he leads the reader, through what he calls interpretation, to the profound wisdom underlying simple notions that are, at face value, false.

Sir Thomas Browne assumed this same corrective posture—without the sympathy for popular sayings—in the next century. In his justly celebrated *Pseudodoxia epidemica: or, Enquiries into very many received tenents, and commonly presumed truths* (London, 1646), he mentions Joubert by name and goes so far as to criticize his *Erreurs populaires*. Browne's use of Bacon's *Advancement of Learning* and familiarity with Descartes's *Discours de la methode* apparently led him to expect from Joubert an error book that entertained not mere physiological problems, but fundamental epistemological questions:

> Laurentius Ioubertus, by the same title [popular errors] led our expectation into thoughts of great releef; whereby notwithstanding we reaped no advantage; it answering scarce at all the promise of the inscription.

But the projects of Joubert and Browne were quite similar: both illustrate a fledgling *episteme* seeking to displace an aging one. Their sources are virtually the same. Indeed, in hollowing out a space for the *Pseudodoxia epidemica*, Browne not only decries the very epistemological stance characterizing his own writings but, in the process, may well have contributed to the English-speaking world's neglect of his French predecessor.

Many other error books were to follow in the tradition launched by Joubert. Gaspard Bachot's *Erreurs populaires touchant la medecine et regime de santé* (Lyons, 1626) offered itself as a fulfillment of a promise, made by Joubert but never kept, to furnish four parts in addition to the two he had published. Jacques Primerose produced a *De vulgi erroribus in medicina* (Amsterdam, 1639), which went through several editions during the seventeenth century (Rotterdam, 1658; Leiden, 1668; [French translation] Lyons, 1689). Primerose, like Browne, faulted Joubert for leaving his work *imperfectum*. Luc d'Iharce's *Erreurs populaires sur la médecine*, citing Joubert's work as paradigmatic, appeared in Paris in 1783. The model proved its ongoing success, for in the nineteenth century Balthasar-Anthelme Richerand authored *Des erreurs populaires relatives à la médecine* (Paris, 1810), followed by a second edition published two years later. Like Primerose and Browne, Richerand claimed that his treatise differed completely from those of his predecessors, but the similarities are far more striking than the differences. Scotland also mined the vein exploited by all of the above: Charles Mackay's *Memoirs of extraordinary popular delusions and the madness of crowds* was published in London in 1869 and continues to enjoy a fair amount of success.

Some readers may well find Book One of Joubert's *Erreurs populaires* tiresome as he dwells on the ethical, sociological, and financial aspects of Renaissance medicine. We have difficulty appreciating his explanations, still firmly grounded in the theory of humors; we fall easily into scorn for the state of medical knowledge at the time. To understand the nature of the discipline in Joubert's day, one might compare the grasp of physiological reality in his time to that of psychopathology today: hypothetical, uncertain, and fraught with conflicting theories, some of which align rather well with certain types of treatment, others that offer satisfying systematic explanations, and yet others that suggest a discipline in crisis. If we allow oblivion to erase the knowledge of the past, or our own epistemological comfort to mock it, we risk losing sight of what that knowledge meant, affectively, to men and women of another epoch. When T. S. Eliot mentioned those readers who feel that "the dead writers are remote from us because we *know* so much more than they did," we must remember his seasoned reply: "Precisely, and they are that which we know." Our own beliefs are inextricably founded on theirs. Their foolishness is none other than our own, wearing not a new mask, but a familiar one.

Students of rhetoric and literature will not be disappointed by the *Erreurs populaires*. The variety is noteworthy: there are powerful epideictic passages, such as those inviting Princess Marguerite de France to accept the dedication of his work, or those enjoining parturient mothers to breast-feed their infants. Joubert's tone borders on anger when he addresses laymen who have little respect for the physician's ability to cure illness, and who misapply information gathered from watching physicians at work or talking among themselves. There are entire chapters that are richly metaphoric. Both the one on the relationship between the physician and the patient's disease and the one on the determination of virginity in a maiden are supported by the language of fortification: last ditch, outer wall, gate, assault, combat, battering ram, gaining ground, breaking in, planting one's banner, and so forth. Joubert juggles metaphors from different registers: military, agricultural, and culinary, all of which were traditionally used obscenely in farce, fabliaux, and facetiae.[4] Moreover, the *Erreurs populaires*, as one might well suspect, contains a host of expressions, adages, and notions reflecting popular beliefs and, in Joubert's eyes, popular misconceptions.

The *Erreurs populaires* gave Joubert the opportunity to express opinions, concerns, and frustrations under the guise of paternally informing, or angrily inveighing against, the relatively defenseless commoners. They are depicted as "the poor, chaste, and continent workers," happy with their lot (Book Two, chapter VII). But other aspects of the culture of the time find expression also, and usually indirectly, such as the custom of kissing all objects before handing them to a prince (Book Two, chapter XIII), or the grim horror of life in a besieged city, one of Joubert's favorite metaphors for a person suffering from an illness. He recounts the valiant efforts of a captain who must resort to having the besieged eat horses, mules, then dogs and cats, and finally the rats of the surrounded city (Book One, chapter IX). But the story of their determination did not end there, and the metaphor takes on life when Joubert cites the place and the date of the siege: Sancerre in the year 1573. The inhabitants ate the leather, the parchment, even "the roof slates, out of which they made bread—I do not know how." Joubert's treatise is a lattice through which sixteenth-century realities can be glimpsed.

Students of philosophy and criticism cannot fail to take delight in the explanations Joubert offers for physiological phenomena and linguistic commonplaces; and those interested in feminist theory will certainly want to take note of what Joubert has to say, both on his own behalf and on behalf of his century, about the primacy of the male sex at a time when men were generally excluded from the practical arena of gynecology and obstetrics (Book Two, chapter IV). The philosophical and theological underpinnings of the *Erreurs populaires* draw constant attention: the theory of humors is applied in numerous specific cases, and the author never seems to tire of reminding the reader that if the physician ever heals, it is always and only with God's help, *Dieu aidant*. Joubert admits the supremacy of the Faculté de Théologie but argues that medicine should precede law in the ordering of the remaining disciplines because the physician's power is over beings and not possessions.

As is the case with many ambitious works, the *Erreurs populaires* was left unfinished. The vast table of projected books reproduced in Appendix A attests to Joubert's energy. Although Part Two of the treatise was published as early as 1580 and appended to Part One with the publication of the Claude Micard edition of 1587, it was Part One that generated violent reaction and led to Joubert's frantic editing in the second edition. Because of the length of Part One (*La premiere partie des erreurs populaires*), Part Two cannot appear in the present volume. The subject matter of Part Two is laid out in Joubert's "Division de toute l'oeuvre" (see Appendix A); an annotated translation of it will be the subject of a second and final volume of Joubert's *Erreurs populaires*.

EDITIONS OF THE *ERREURS POPULAIRES*

The first extant edition of the work bears the following title: *Erreurs populaires au fait de la medecine et regime de santé, corrigés par M. Laur. Joubert Conselher & medecin ordinaire du Roy & du Roy de Nauarre, premier docteur regent, Chancelier et iuge de l'uniuersité an medecine de Mompelier. Cette-cy est de toute l'oeuure la premiere partie, contenant cinq liures, auec l'indice des matieres qui seront traitees ez autres.* It was published simultaneously in Bordeaux, Paris, and Avignon in 1578. If we are to believe Simon Millanges, the Bordeaux printer responsible for Montaigne's *Essais*, he was in legal possession of exclusive printing rights for Part One. Millanges produced in 1579 the hastily revised second edition of Part One. Any reading or translation should, therefore, take into account at least these two editions. The first edition was done with extreme care and, for reasons which will be discussed shortly, served as the basis for the scores of editions and printings, both legal and pirated, of the *Erreurs populaires*. Since Joubert died in October of 1582, it is safe to assume that any differences existing between these two editions taken together and other contemporary or later editions are not attributable to Joubert. The Gordian knot of collating the nineteen known editions of the *Erreurs populaires* is thus cut, and the principal task becomes that of comparing the first Millanges edition with the second. Their differences are considerable.

The following is a table of all known editions and at least one location in which a copy is held:

1. Bordeaux, 1570 [mentioned by Amoreux]: as yet not found.
2. Bordeaux, 1578 [Millanges]: Bibliothèque Nationale and the Bibliothèque Municipale de Montpellier.
3. Avignon, 1578 [Roux]: Bibliothèque Municipale de Lyon.
4. Paris, 1578 [V. de Mehubert]: Bibliothèque Nationale.
5. Paris, 1579 [Part Two, L'Angelier]: Bibliothèque Municipale de Montpellier.
6. Bordeaux, 1579 [Millanges]: Bibliothèque Nationale; Bibliothèque Mazarine; Bibliothèque Municipale de Bordeaux; Surgeon General's Library (Washington, D.C.).
7. Paris, 1580 [Part Two, L. Breyer]: Bibliothèque Nationale.
8. Bordeaux, 1584 [Millanges]: New York Surgical Library.
9. Avignon, 1586 [Roux]: Bibliothèque Nationale.
10. Paris, 1587 [Parts One and Two, Claude Micard]: Bibliothèque Nationale; Bibliothèque Municipale de Lyon; Bibliothèque Mazarine; New York Academy of Medicine.
11. Florence, 1592 [Italian translation, F. Giunti]: Bibliothèque Nationale; Bibliothèque Mazarine.
12. Antwerp, 1600 [Latin translation, M. Nutii]: Bibliothèque Nationale; Bibliothèque Mazarine.
13. Rouen, 1601 [Parts One and Two, Raphaël Du Petit Val]: Bibliothèque Nationale.
14. Rouen, 1601 [Parts One and Two, Pierre Calles]: Bibliothèque Nationale.
15. Paris, 1601 [Parts One and Two, unnamed printer]: Bibliothèque Municipale de Lyon.
16. Rouen, 1601 [Parts One and Two, T. Reinstadt]: Bibliothèque Municipale de Bordeaux.
17. Lyon, 1602 [Parts One and Two, Pierre Rigaud]: Bibliothèque Municipale de Montpellier.
18. Lyon, 1608 [Part Two, Pierre Rigaud]: Bibliothèque Nationale.
19. Lyon, 1608 [Parts One and Two, Pierre Rigaud]: Bibliothèque Nationale; Bibliothèque Mazarine.

This listing does not account for reprintings, the number of which is difficult to determine, but it is safe to assume that they were numerous (Cottrell claims there were ten in only the first six months), although not as widespread as the recipe and remedy books of which Natalie Zemon Davis speaks. Luc d'Iharce, who wrote two centuries after Joubert in the same popular error tradition, claims in his own *Erreurs populaires sur la médecine* (Paris, 1783) that Joubert's work was so immensely in vogue that it was printed "in four different places, namely Bordeaux, Paris, Lyons, and Avignon, and that in each place not less than sixteen hundred were run off." Nearly all these editions and reprintings are based on the 1578 edition, and not the 1579 edition, which contains numerous added chapters and passages, mostly in response to the criticism launched by Sainte-Marthe and the midwives against the first edition.

The first edition was chosen by the various printers for at least two reasons. Since it provoked such a violent reaction, it promised to sell better than one modified to satisfy various censuring bodies. Second, from a technical point of view, the 1578 Millanges edition was very carefully done in a large font, with typographical errors accounted for in an errata page bound at the end. One of the most telling proofs of the paternity of the 1578 edition is that Joubert's son, Isaac, sent Barthélemy Cabrol four chapters of the *Erreurs populaires*, along with a gathering of popular adages (mostly the work of Jean Momin, a doctor of medicine at the university of Montpellier), thinking they were unpublished drafts from Part One. The fact that this material, which appeared in the 1579 edition, escaped the notice of Joubert's son is a striking indication of that edition's weak currency. Furthermore, when Cabrol, to whom Joubert gave permission to publish Part Two, was later assembling updated materials for a definitive version (the 1587 Micard edition), we see that he not only based it on the 1578 Millanges edition for the most part, but that he also included the four chapters Isaac had sent him. Neither Isaac Joubert nor Cabrol had carefully read the revised 1579 Millanges edition (see note 8 of chapter VIII in Book Four).

INVENTORY OF THE 1578 MILLANGES EDITION

1. Dedicatory letter to "Marguerite de Navarre" [Marguerite de France; in the present translation this letter immediately precedes the five books of the treatise, as explained in my Introduction; see also note 3 of the Dedicatory Letter to Marguerite de France].
2. Table of contents ("Division of the Entire Work into Six Parts Containing Thirty Books") promising an eventual 309 chapters [see Appendix A].
3. Medley ["Melange": a sundry gathering of popular sayings, notions, or fragments of arguments serving as a basis for the *Erreurs populaires*; seventy items; appearing in front matter of the present translation].
4. Letter by Joubert to the reader.
5. Nine liminal poems [see Appendix B].
6. Books One through Five.

7. Brief apologetic letter to Marguerite de Navarre [Marguerite de France; see Appendix D].
8. A short treatise appended to many of Joubert's works, the *Question vulgaire* ["What language would a child speak if it had never heard speech?"; to appear in volume two].
9. Brief letter from the printer [Millanges] to the reader [see Appendix C].
10. Errata [one page; not reproduced or translated herein].
11. Copy of privilege and right to exclusive production of Part One [see Appendix E].

DESCRIPTION OF THE 1579 MILLANGES EDITION

The 1579 edition was prepared in great haste. It suffers not only from excisions dictated in large part by vehement criticism, but also from additions inserted with little concern for coherence. Typographical errors abound, and there is no errata page. It is clear that Joubert worked feverishly, as did Millanges, to reshape a work that had rocked the medical world, at least momentarily, only to find a few years later that the original, causing so much scandal, was to become the most sought after and, consequently, the standard edition of the *Erreurs populaires*.

The treatise, as it appeared in its reworked form in 1579, begins with a letter by Louys Bertrauan defending Joubert against the "envenomed" attacks occasioned by the first edition. Bertrauan claims that a printer's error was the source of the obscene word *vit*, which should have read *vir* (see note 1 of the Dedicatory Letter to Marguerite de France). He then excuses Joubert for the graphic terms contained in the midwives' depositions reproduced in chapter IV of Book Five, stating that these are legal documents and that the language in them cannot, therefore, be cause for shame. His apology reminds readers that canon law and even the Bible (e.g., Genesis 19 and Leviticus 18) contain passages that Joubert's critics would doubtless find objectionable. Bertrauan pleads for an unexpurgated *Erreurs populaires*, using the metaphor of game offered to a lord or lady in its entirety, and not eviscerated (*esventrée*).

Next comes a letter written by Joubert to his friends, assuming the same defensive posture and claiming the right to designate things by the "proper" name given them by divine inspiration. He notes that man accords eyesight more delicate treatment than hearing, "For will one not name with infinitely less shame the bottom (speaking reverently) than one will show it." According to Joubert, proper terms in themselves do not "reek" but are "good and legitimate" for expressing "properly what we mean." He follows Bertrauan's lead in denouncing the printer's error (*vit* for *vir*) and then defends the vocabulary of his subject matter as inoffensive to any married woman, although admittedly improper for the unmarried. Joubert justifies his treatment of what were considered evil practices (the sin of Onan, for example) on the grounds that they must be named if they are to be decried publicly, citing as predecessors in this enterprise Plutarch, several Roman emperors and legislators, Saint Paul, and other authors of the Bible. He closes his letter with a blanket apology to Marguerite and "to all who were offended by [his] discussion," maintaining that his intention was exactly the opposite.[5]

Joubert's dedicatory letter follows. It is the one addressed to Marguerite in the 1578 edition but replaces the name of Marguerite with that of Guy du Faur de Pibrac. Because it begins the argument proper of the *Erreurs populaires*, I have placed it immediately before the five books of the treatise. In the 1579 edition, however, it is followed by the same table of contents appearing in the 1578 edition, except for the added chapters in Books One and Two. Only three items remain in the Preliminary Matter: (1) the medley ("Melange"), considerably expanded by sixty-two new items; (2) the letter by Joubert to the reader; (3) twelve liminal poems (three having been added to the nine of the first edition); and (4) a brief letter to the reader by the printer, somewhat different from that appearing in the first edition, as can be seen from translations of them in Appendix C.

After the five books of the treatise come the following items: (1) *Question vulgaire*; (2) three texts written by Joubert but furnished by a certain A. I. D. N., "Bachelier an medecine de l'université de Mompelier," (a) "Du breuuage de Monseigneur le Marechal d'Anville" [a brief tract discounting the substitution of wine with a concoction invented by Anville], (b) "La santé du prince",[6] (c) "Du serain" [a discussion of nightfall and the possible dangers accompanying it]; (3) a letter from the printer, Millanges, to the reader; (4) an extract from the "privilege" to print the *Erreurs populaires*. Because the above texts—except for those items noted as translated herein—were incorporated into Part Two of the treatise, they will appear in the second volume of my translation.

TABLE OF EQUIVALENCE

The following table indicates instances of equivalence between chapters in the 1578 edition and those in the 1579 edition. Those chapters appearing for the first time in the 1579 edition, as well as those that underwent modification, are marked with an asterisk. For example, because of the insertion of two new chapters into Book One of the 1579 edition, what was chapter VII in the 1578 edition became chapter IX in the 1579 edition, and no changes were made therein.

1578 MILLANGES EDITION		1579 MILLANGES EDITION
Book One		
I	=	I
ii	=	ii
iii	=	iii
iv	=	iv
v	=	v
vi	=	vi
. . .		vii*
. . .		viii*
vii	=	ix
viii	=	x
ix	=	xi
x	=	xii
xi	=	xiii
xii	=	xiv
xiii	=	xv
. . .		XVI*
XIV	=	XVII
xv	=	xviii
xvi	=	xix
. . .		xx*
Book Two		
i	=	i
ii	=	ii*
iii	=	iii
iv	=	iv
v	=	v
vi	=	vi
vii	=	vii*
viii	=	viii*
ix	=	ix
x	=	x*
xi	=	xi*
xii	=	xii
. . .		xiii*

1578 MILLANGES EDITION		1579 MILLANGES EDITION
Book Three		
i	=	i*
ii	=	ii*
iii	=	iii*
iv	=	iv*
v	=	v
vi	=	vi
vii	=	vii*
VIII	=	VIII
ix	=	ix
Book Four		
i	=	i
ii	=	ii
iii	=	iii
iv	=	iv
v	=	v
vi	=	vi
vii	=	vii*
viii	=	viii*
ix	=	ix
x	=	x
xi	=	xi
xii	=	xii
Book Five		
i	=	i
ii	=	ii*
iii	=	iii*
iv	=	iv*
v	=	v
vi	=	vi
vii	=	vii
viii	=	viii
ix	=	ix*
x	=	x
xi	=	xi*

In all cases, specific deletions, additions, changes in marginal notation, and typographical errors of consequence are pointed out in the notes. I have used only one set of endnotes. In the case of a marginal note by Joubert, I have used a footnote where it appears in the text of the translation and placed it in the endnotes with the indication "Joubert's note"; the text that follows the colon is a translation of the marginal note. If comments are added to the note, they are set off in a new paragraph.

SELECTED BIBLIOGRAPHY

All ancient texts cited herein are from the Loeb Classical Library unless otherwise indicated.

Amoreux, Pierre-Joseph. *Notice historique et bibliographique sur la vie et les ouvrages de Laurent Joubert, chancelier en l'Université de médecine de Montpellier, au XVIe siècle.* Montpellier, 1814.

Bowen, Barbara C. *Les Caractéristiques essentielles de la farce française et leur survivance dans les années 1550–1620.* Urbana, 1964.

Brabant, H. *Médecins, malades et maladies de la Renaissance.* Bruxelles, 1966.

Broussonnet, J. L. V. *Notice sur Laurent Joubert, professeur et chancelier de l'Université de médecine de Montpellier.* Montpellier, 1829.

Cotgrave, Randle. *A Dictionarie of the French and English Tongues.* London, 1611.

Cottrell, Robert. *Sexuality/Textuality: A Study of the Fabric of Montaigne's 'Essais.'* Columbus, Ohio, 1981.

———. "Une source possible de Montaigne: le *Traité du ris* de Laurent Joubert." *Bulletin de la Société des Amis de Montaigne* 9–10 (1982), 73–79.

Davis, Natalie Zemon. *Society and Culture in Early Modern France.* Stanford, California, 1975.

Dulieu, Louis. "Laurent Joubert, chancellier de Montpellier." *Bibliothèque d'Humanisme et Renaissance* 31 (1969), 139–67.

Eccles, Audrey. *Obstetrics and Gynaecology in Tudor and Stuart England.* Kent, Ohio, 1982.

Gordon, Benjamin Lee. *Medieval and Renaissance Medicine.* New York, 1959.

Gourg, Jean-Louis. "Laurent Joubert, le *Traité du ris* et Rabelais." In *Mélanges de philologie offerts à Charles Camproux,* 679–87. 2 vols. Montpellier, 1978 .

Huguet, Edmond. *Dictionnaire de la langue française du seizième siècle.* Paris, 1925–67.

Hurd-Mead, K. C. *A History of Women in Medicine from the Earliest Times to the Beginning of the Nineteenth Century.* Haddam, Connecticut, 1938.

Longeon, Claude. "Louys Papon, Laurent Joubert et le *Traité du ris*." *Réforme Humanisme Renaissance* 7 (1978), 9–11.

Maclean, Ian. *The Renaissance Notion of Woman: A Study in the Fortunes of Scholasticism and Medical Science in European Intellectual Life*. Cambridge, 1980.

Montaigne, Michel de. *Oeuvres complètes*. Ed. Maurice Rat. Paris, 1962.

Nicaise, E. *'La grande chirurgie' de Gui de Chauliac*. Paris, 1890.

Oxford English Dictionary. Compact Edition. Oxford, 1971.

Pauly, August Friedrich von, and Georg Wissowa. *Real-Encyclopädie der classischen Altertumswissenschaft*. Stuttgart, 1894–1963. And Supplement, 1903–78.

Rabelais, François. *Oeuvres complètes*. 2 vols. Ed. Pierre Jourda. Paris, 1962.

Rocher, Gregory de. "Le rire au temps de la Renaissance: le *Traité du ris* de Laurent Joubert." *Revue Belge de Philologie et d'Histoire* 56 (1978), 629–40.

———. *Rabelais's Laughers and Joubert's 'Traité du ris.'* University, Alabama, 1979.

———, trans. *Treatise on Laughter*, by Laurent Joubert. University, Alabama, 1980.

———. "Quelques précisions sur l'oeuvre de Laurent Joubert. *Bibliothèque d'Humanisme et Renaissance* 43 (1981), 345–46.

———. "Note sur le mot *destil*." *Bibliothèque d'Humanisme et Renaissance* 44 (1982), 615–16.

———. "The Trouble with Women: Some Medical Musings from Sixteenth-Century France." *Renaissance Papers 1987* (1988), 39–47.

Sabatier, Robert. *La poésie du seizième siècle*. In *Histoire de la poésie française*. Paris, 1975.

Screech, M. A. *The Rabelaisian Marriage: Aspects of Rabelais's Religion, Ethics, & Comic Philosophy*. London, 1958.

———. *Rabelais*. Ithaca, New York, 1979.

Smith, Wesley D. *The Hippocratic Tradition*. Ithaca and London, 1979.

Wickersheimer, E. "Un brave homme et un bon livre: Laurent Joubert et les *Erreurs populaires au fait de la médecine et du régime de santé*." In *La médecine et les médecins français à l'époque de la Renaissance*. Paris, 1906.

POPULAR ERRORS

LAURENT JOUBERT

LOUYS BERTRAUAN, DOCTOR OF MEDICINE TO ALL GREAT LOVERS OF VIRTUE, SALUTATIONS

Messieurs, as soon as my good friend Monsieur Joubert had dedicated to the most illustrious queen of Navarre Part One of the *Erreurs populaires* he well suspected that he would be calumniated on account of it. He therefore thought he would immediately attach his apology to Her Majesty at the end. He does this in an even more fitting manner in this second edition, which has been revised and corrected.

This should suffice for those who found such a dedication inappropriate due to the few earthy terms he is forced to use in his arguments. These words would indeed be improper if addressed to a maiden who ought not to know of such things; but every married woman can in all decency read and understand everything that is contained in this work, not any less than in the *Heptameron or Stories of Fortunate Lovers*, tales by the most illustrious and most excellent Princess Marguerite de Valois, former queen of Navarre, dedicated to the late queen, her daughter.[1]

For there are tales that are just as bawdy (as they say) as those of Monsieur Joubert. Still, there is this difference, in order to excuse all the more our author: he is a philosopher and a physician who speaks of natural things and the marvels of God, in which matters he is obliged to explain and uncover with decency several secret things. The other author is a princess, dealing with love stories in a carefree way, which must not, by the way, be taken badly by any means. And what word (I ask you) in this entire book do you find that ought to be considered foul and dirty, unless perhaps the one on page 213 in the 23rd line from the top of the page where there is a *t* in the place of an *r*? For it should read *vir*, as it does in its corrected form.[2] As for the other words in the three depositions, they are not any more scandalous: they are heard so often on everybody's lips that they can be said openly, and that is all there is to be said on the matter.

And it must not be believed, as some grumpy detractors and deceivers do, that Monsieur Joubert found and forged these depositions; for Monsieur de La Valade, master of requests of the king of Navarre, a decent and upright man, will always certify having handed over to him the one from Béarn and the one from Paris, both of which he considers to have been duly sworn. And can acts of justice be considered indecent?

"But there are, in the chapter on the reckoning of virginity in a maiden, certain words that are not edifying for girls," someone will say. I answer that complaint by saying that Monsieur Joubert, in this second edition, begs girls not to look at this chapter, which women are of course allowed to read. But even so, what does he say that is so scandalous? Does he not decently abhor in this chapter that which a few crazed women do with a dildo, as it is called, a well-known and most notorious instrument?

In order to tax vices and sins one is forced to uncover and describe them. Do you call that instructing in evil? It would then be necessary to condemn Holy Writ, which mentions numerous illicit couplings between parents and kinsmen, even males with males, and (what is most horribly abominable) men and women with animals, as Moses gives us to understand in detail. How many people are there in the world who know nothing of all this, and is there even one person who simply cannot think that these things have been done? One therefore instructs such people in evildoing? Far from it. But one is forced to speak out against those who commit such abominations.

In the holy volumes of canon law, by which the greater part of Christianity is governed, we learn by reading them (if I dare speak in such a manner) of a thousand filthy practices previously unheard of by many. Must one therefore abstain from reading them? Are they kept from decent people? Moreover, will a decent woman or maiden not dare to read or hear read what is recounted in the nineteenth chapter of Genesis or in the eighteenth chapter of Leviticus, and in so many other passages of Holy Writ?

One could raise to Monsieur Joubert the objection that it would have been advisable to leave aside the chapters which treat of disgusting matters and dedicate only the rest to Her Majesty, and at least not ask her to serve as judge in the matter. I shall reply to you: such had always been his intention, but having communicated it to the principal people at the court of King Henry of Navarre, his master, he was given to believe that it could all be published together, especially since he had dedicated to Her Ladyship the entire work of the *Erreurs populaires*, which will contain at least thirty books, already sketched and laid out. For among the three or four hundred chapters that are to be, three or four could not be taken out because, had this been done, it would have been necessary to start the typesetting all over again. Someone put it very gracefully: when people offer game, a roebuck, a hare, or some other animal to a lord or a lady, they offer it in its entirety, without having gutted it. Does that mean one is therefore offering dung? The same is true with a written work, in which there is good and bad, both left to the judgment of the reader. Take what you find to be good, leave the rest; another will surely want what you reject.

But I know well what Monsieur Joubert says to those who speak to him now about this matter, after the fact: that if Her Majesty had not enjoyed the work he would have heard about it from a large number of friends who are close to her. And if his offering is pleasing to the saint to whom it was made, everyone should be happy over it; and if otherwise, he will take it back, as should be the case if it is not well received, and say openly: "I wish I had never written it."

Farewell.

THE AUTHOR TO HIS FRIENDS AND TO THOSE WHO SPEAK WELL OF HIM, SALUTATIONS AND BLESSINGS

How well did I foresee that I would be calumniated because of a few words contained in my *Erreurs populaires*, and especially because I dedicated it to the most illustrious and excellent queen of Navarre, one of the most chaste and virtuous princesses in the world. As if the explanation of natural matters such as conception, pregnancy, birth, and lying-in—all things that people wish to know about and that they seek out every day, men as well as decent women—ought not to be presented to such a lady and true model of virtue.

I had indeed predicted this, and I debated with myself for quite some time (but for too long, to my great regret). The criticism came after I had already revised and published my work to comply with the judgment and censuring of all parties, great and small, learned and ignorant, friends and enemies, benign and malicious, those who speak well and those who speak ill of me. Nevertheless, I think I have written rather modestly, considering the subject (the organs and functions that decency orders us to keep covered and hidden), speaking of them in a similarly covered and masked manner, in disguised words. Furthermore, I did not use words in their literal sense;

such words are not in my vocabulary, although the Romans and the Greeks, and perhaps the Hebrews, in their most decent books do not shy away from them. For one is not ashamed to speak literally of all the parts of our bodies in Hebrew, in Greek, and in Latin, according to the names that were first assigned to them. This has, as one might suppose, been the case with whores and lascivious people, bawds or panders, and other filthy, infamous, profane, and shameless people, but also with the wisest and most prudent onomatologists (that is, authors and givers of names), of whom the first and most worthy of such a charge was our grandfather, Adam, inspired by God in the knowledge of all things, which he designated so properly that their names corresponded to their essences.

The most wise and divine Plato in his *Cratylus* (which is about the true source of names) points out with subtle prudence that not just anyone can impose the appropriate name on something in nature, but only the most excellent personages, who know well how to explain things through the name each carries. Lascivious people, haunters of whorehouses, idle streetwalkers, tavern keepers, and other similar rascals and scum make up a few throaty and obscene words (as they say) or apply normal words in an extravagant manner to mean something else for some merriment or nastiness, and the beggars use their own gibberish, which aids in their deception.

But the true and literal names of all things, down to the smallest and the most hidden, are handed down from the learned ancients and philosophers, inspired by God, as was Adam. These were also the ones who taught men to live in a human, decent, and civil manner. This is why Plato says, and rightly so, that the assigning of names comes from a mind that is more divine than human. Hence, literal terms are decent in any language, provided they are used decently and to the point, in order to explain the subject being treated. Otherwise, those good people who assigned them in Hebrew, Greek, and Latin would have been gravely mistaken and should rather have taught us how to use periphrases and circumlocutions, which have since been invented in order to speak more secretly about that which we nevertheless wish to have clearly understood when we designate what we are ashamed to look at.

Or would it have been better if the learned ancients had never assigned names to things that should be kept secret and hidden, so that they could not be spoken of at all? But what would it be like if we had a familiar acquaintance with things and yet did not know their names? Would it be more decent to point at them to make people understand what we want, just as the dumb point at that which they cannot say or name? It is certain that we have always held our eyes in more respect and honor than our ears, for will one not name with infinitely less shame the bottom (speaking reverently) than one will show it?

Thus, literal terms (as is said in the common proverb) do not reek but are themselves good and legitimate, so we can decently use them to designate effectively and properly what we mean, just as do the most modest and virtuous people in all languages, even more so in writing than in speaking familiarly. For, truly, the license is greater in writing, since one writes to people of every station, even though the dedication or the inscription is to some great person, commonly called a demigod.

Nevertheless, I abstained from using literal terms for the shameful parts. The foul word on page 213 is not mine, but a corrupted word for *vir*. Such expressions have never been in my vocabulary, even though I admit to joking during my public dissections and to speaking freely and lightly about these very parts when the subject inspires me. But I call as witnesses the thousands upon thousands of listeners I have had over the years—the physicians, surgeons, and apothecaries who are now all over Europe—if they have ever once heard me pronounce a literal term for the shameful parts or for the venereal act.

It is necessary to come now to my subject, covered and disguised under common words, and to examine whether it is so lascivious and indecent that proper women could not read it or have it read to them. And where is it, I ask you, that I have written about indecent and scandalous matters, inciting women or girls to do otherwise than decency and virtue require? If I touch upon the subject of the flesh, it is that between a husband and his wife, concerning conjugal love, marriage, conception, pregnancy, birth, and lying-in. Or if I happen upon lasciviousness, it is to decry it every bit as severely as a theologian would.

The acts that my words designate can only be understood by women who are legitimately allowed (because of holy matrimony) to have a certain knowledge and enjoyment of such things. Girls can learn nothing from this, neither good nor evil, if they have not been instructed in these matters elsewhere, as (unfortunately) most of them have already been, and too well. But it has never been through my words, whether public or private (and still less through my writings), that they have understood anything. It is rather the bad company. It is the too-familiar conversations with debauched, incontinent, and lascivious men, too free with their words and their hands, along with the reading of books about love (be they in poetry or prose), tales (either from history or from fables) about the wicked tricks wives play on their husbands, and, vice versa, boldly committing adultery (which should be no less punished than publicly condemned).

It is language like this, not mine, that corrupts the chaste hearts of women and girls, so help me God. It is against such language that one should inveigh, because of the scandal it brings, and the evil example, and the stirrings of the flesh, always too inclined to go along. Or at least forbid girls to read such books (which they are permitted to read altogether too much). One can only object here to my chapter on virginity. But stop to think! Will girls learn from it how to do evil when I am uncovering for a good reason the folly of a few who are unchaste of heart and have a penchant toward lasciviousness (as I say in that particular chapter)? That reason is to discourage the good and the modest from such practices, so they will not pollute themselves in such a foul manner and corrupt their mark of maidenhood, more valuable than all the wealth in the world. And do you think that I would have written about such a matter if I did not know for a fact that there are girls who do something far worse? But I kept from speaking about it because it is less widely known.

I have seen what Master Louys Bertrauan (who, I add, is very knowledgeable about my writings) says in my favor, raising the point that there are some passages in Holy Writ where extreme cases of lasciviousness are recounted, and the point that several decent and simple people, men and women, have never heard of such things. And it is true, as he points out also, that wise emperors and Roman legislators mention and censure in their laws a few abominable and corrupt acts of the kind we would choose not to speak about.[3]

But since corruption is so widespread that men and women are worse than animals, one is forced to loathe and decry it publicly. Do the cases of incest mentioned in the Holy Bible, the incredible infamies touched upon by Saint Paul writing to the Romans, and the villainy of Onan, who spilled his seed on the ground, constitute a scandal for the God-fearing reader, apprising him or inducing him to commit evil deeds? May it please God that this is not so![4]

One esteems very highly the reading of the moral works of Plutarch, and for a good reason, for he is a serious author with much that is good and great to teach us. Now, in the tract he wrote called "Love and the Natural Affection Fathers Have for Their Children" [*Peri tes eis ta ekgona philostorgias*], he becomes rather direct in his lusty comments so as to push women and maidens into love's games. But in the tract called "Love" [*Peri philadelphias*], in which Bachon (nicknamed the handsome son) is sent out into the country and given a discourse on conjugal love and sodomy (as we call it), there is a great deal of filth added gratuitously. I am speaking of the content (for the terms are most decent), because when one wishes to refer to sodomy, even if for the purpose of condemning it, one cannot speak of such a horrible vice without considerable scandal.

This is one thing I could say in my defense against those who are offended by my subject and the dedication of my work. I truly think they believe that I am unaware of having blundered (for who is so bold and senseless as not to be terrified and horrified at offending such a good, virtuous, and great princess?), but that I acted out of ignorance, absentmindedness, and inadvertency, as one often does when one writes a great deal. For, as the wise man says, in such a great number of words it is impossible for there not to be some bad ones. Just as the noble Homer on occasion nods and dreams, so I, as in my public dissections, thought perhaps I was addressing my students. All these things can be said, thought, and believed by those who wish me well and who are kind enough to have a sound opinion of me.

But I am not about to accept that and wish, rather, to change the dedication so that none will be scandalized over the person whom I will set up as judge of this work. I would have liked to have called back a good part of my treatise, but it was so quickly disseminated that it was impossible for me to recall it to withdraw certain parts in the three or four thousand copies that had been released. I would especially liked to have removed the midwives' depositions, more inept than indecent (because the terms used in them are unknown), and other parts which might have caused displeasure to a minority in France. I would thus have been in accordance with the doctrine of Paul the Apostle, which states that everything that is permissible is not expedient, and that one ought to abstain from what is otherwise permitted if some people might be scandalized by it.[5]

Begging most humbly the most illustrious, most august, most wise and magnanimous queen of Navarre, and all those who were offended by such sections, to excuse my enterprise, the purpose of which had been quite the opposite.

TABLE OF CONTENTS OF THE *POPULAR ERRORS*

THE FIRST BOOK OF POPULAR ERRORS CONCERNING MEDICINE AND PHYSICIANS

THE SECOND BOOK OF POPULAR ERRORS CONCERNING THE VENEREAL ACT, CONCEPTION, AND GENERATION

THE THIRD BOOK OF POPULAR ERRORS CONCERNING PREGNANCY

THE FOURTH BOOK
OF POPULAR ERRORS CONCERNING
CHILDBIRTH AND LYING-IN

THE FIFTH BOOK
OF POPULAR ERRORS CONCERNING
MILK AND THE CHILD'S NOURISHMENT

MEDLEY [1578]

---

1. Why it is said that many marriages are unlucky.

2. Whether it is right to say that a pale woman is in need of a male.

3. Whether it is true that men become older for having gone to bed with an old woman, and that old women become younger for having gone to bed with a young man.

4. Casting a spell of impotence; and how it is done.

5. Whence comes it that girls normally speak sooner than boys.

6. Against those who think that by taking out a man's milt he will be made the nimbler.

7. Whether it is true that garters stunt growth and cause wrinkles in girls.

8. Concerning hermaphrodites, otherwise called fainthearted milksops; and whether it is possible that a woman become a man, or vice versa.

9. Why it is said that when someone has a nosebleed he will soon hear some good news.

10. Whether it is true that a dying person will suffer more if there is a partridge feather in his bolster or pillow.

11. Whether it is true that a child has one-half the height he will attain at the age of three.
12. Whether it is true that biting one's fingernails causes nearsightedness, as some people say.
13. Why one says to children who handle fire, or carry it throughout the house, that they will wet the bed.
14. Why it is said of him who is brusque and lusty that he is a monstrous cutter or a terrible gallant.
15. Against those who do not want ailing nipples to be touched either by salve or by iron.
16. Why people say that a good cold lasts forty days.
17. Whether it is right to say that a pear with cheese is bound to please [*La poire avec le fromage est mariage*].
18. Whether it is true that a turquoise received from a friend without having asked for it preserves from injury by fall, if the stone breaks.
19. Whether wearing an amethyst keeps one from getting drunk.
20. Why people say yawning cannot deceive: it means one wishes to eat, sleep, or quit one's loves.
21. Whether it is true that a tonsured man has less strength.
22. Why it is considered healthy to fart while pissing.
23. Whether it is true that from the scurf one has on the wrist one can tell that there is also some on the buttocks.
24. How it is that from a salty forehead one can tell that a child has worms; and what the most definite symptoms of vermin are.
25. Whether it is good to stop children from using their left hands.
26. Why people say there is no sauce like appetite; and whether it is good to use sauces occasionally.
27. Whence snoring comes; and whether having the head low or sleeping on one's stomach can cause it.
28. Whether one can stop another from snoring by putting his slipper, shoe, or boot under his bolster.
29. Whether it is true that sleeping with the head low causes dreaming; and whether eating cabbage causes it also.
30. Why common people say he who hasn't a hard tummy won't sleep very soundly.
31. Whether it is true what they also say: *Ionture non vau onchure,* "Joining does not beat oiling," or "Coupling does not equal greasing."
32. Against those of the opinion that surgeons are not fit to reset dislocations; and who believe empirical resetters would be more efficient.
33. Concerning those who hate certain foods such as bread, wine, cheese, apples, game, etc.; and whether it is a good or bad inclination.
34. Concerning those who are able to go without drink for two or

three months and more, and others who go longer without drinking or eating.
35. Why people say that one does not grow older at table or at Mass.
36. Whether it is true that one grows as long as one sleeps; and that the day's work takes as much from one's height as is gained while sleeping.
37. Why people say that an unexperienced physician fattens the churchyard, and that bad physicians arrive on horseback and leave on foot.
38. Whether it is true that trouble comes by the pound and goes away by the ounce, or that it comes back posthaste and leaves lackadaisically.
39. How the patient is reproached by the doctor; and how he is blamed in particular for all his excesses and defects.
40. Whether it is true that a woman will not conceive or retain semen if she urinates soon after copulation.
41. Whether it is true that ruptured or burst men make more children than others.
42. Why people say that a satiated woman is as good as pregnant.
43. Whether it is true that nocturnal emissions would be so many children.
44. Concerning pregnant women who drink aquavit as soon as they enter their ninth month so that their child will not be born scurfy.
45. Concerning those women who do not want somebody to go for fire in the house of a woman who just gave birth, for fear that the child will foam at the mouth or be blear-eyed.
46. Whether it helps the delivery for a woman to say three times (while quickly wiggling the thumb), "I'm cold, I'm hot."
47. Whether, in order to give strength to a severely underweight child, the change to stale milk is necessary.
48. How it can be that an absent wet nurse knows in her nipples that her child is crying.
49. Whether it is true that a child will have its buttocks scalded if one throws glowing coals or hot ashes on its excrement.
50. What the symbolism is of the gift of eggs and salt to a friend's foster child the first time one comes to his house.
51. Whether the ends of the fingers being fat is a sign that the person will become fat; and whether slim fingers are a sign of leanness.
52. Against those who say we will live until we die in spite of physicians.
53. Whether it is true that by kissing children often we absorb their blood.
54. Whether it is correct to say that one must drink between cheese and pear.

55. Why people say: "After an apple, no man ever drank," and "After a pear, wine or the priest."
56. Whether it is true that apples, pears, and nuts ruin the voice.
57. Whether it is right to say that milk with fish is poison, and after fish, nuts are an antidote.
58. Why people say about young meat and old fish that the meat brews fat and the fish poison.
59. Against those who say: "Piss clear and make a fig at your physician"; and against others who say: "He who sleeps, pisses, and jiggles well has no need of Doctor Bell." Also: "He who has sanicle and bugleweed can thumb his nose at doctors."
60. That the man who thinks himself in good health carries his death in his bosom; and whether it is true what one says: "Far from a city, far from health."
61. Whether it is madness to make one's physician the beneficiary of one's will.
62. Whether it is right to say: "Against death the true shield is bread and cheese," and "All cheeses are wholesome if they come from a miser."
63. Why people say: "A young man who abstains from sleep and an old man who indulges in it both are on the road to death," and "Who goes to bed late and rises early will soon see his end."
64. Why people say that a joyous heart makes a gorgeous face.
65. Whether it is right to say that he who wishes to be old later must take measures early, and that he who wishes to be in good health must let himself die of hunger.
66. Why people say: "He who drinks no wine after *salade* is in danger of becoming *malade.*"
67. Whether it is true to say that a headache wants to eat and a bellyache wants to shit.
68. Why people say: "A toothache is a pain from our parents," and "A side ache is a stone in the field."
69. Why a woman and a pear without rumors are held in such esteem and value.
70. Whether it is right to say: "The top, the bottom, and the middle hot, the rest is of no consequence."

MEDLEY [1579]

1\. = 5. [1578]
2\. = 6. [1578]
3\. = 8. [1578]
4\. = 9. [1578]
5\. = 12. [1578]
6\. = 13. [1578]

7. If there is any foundation for saying somebody is talking about a person whose ears perk up.

8. Foolish superstition of not biting one's nails on days that have an *r* in them, but that one must closely observe the moon in this matter, as well as in cutting one's hair.

9\. = 18. [1578]
10\. = 19. [1578]
11\. = 20. [1578]
12\. = 21. [1578]
13\. = 23. [1578]
14\. = 24. [1578]
15\. = 38. [1578]
16\. = 39. [1578]

17. = 64. [1578]
18. = 65. [1578]
19. = 67. [1578]
20. = 68. [1578]
21. Which is more inebriating, old wine or new wine.
22. Whence comes it that he who is drunk gets all the more drunk when he is put by a window.
23. How one can make a person who abuses wine hate it.
24. = 29. [1578]
25. = 63. [1578]
26. = 63. [1578]
27. = 30. [1578]
28. = 27. [1578]
29. = 28. [1578]
30. Whether or not snoring is a sign of health, as people say.
31. How it is that pleasant odors and sweet things move the womb.
32. = 16. [1578]
33. Concerning the sick people made to walk in the streets with tambourines and songs to keep them from sleeping.
34. = 22. [1578]
35. Superstitious and foolish error of those who think that if one is well fed at table, one of the guests will die within a year.
36. Error of those who say that an ant found crawling on a person foretells a winding sheet.
37. = 10. [1578]
38. Whether it is possible to know the day and the hour of death.
39. Whether watered-down wine slakes thirst better than pure wine.
40. Why salad is eaten more often at supper than at dinner.
41. Why exercise is better before a meal than after.
42. Why people say: "In the morning, mountains; in the evening, fountains."
43. Why people say that wine is milk for old people.
44. Why all pains are usually greater at night than during the day.
45. Why people say that from studying too much one goes crazy, as affirmed by the second-born among the dead, who had bad luck because of it. [?]
46. If it is true that people on a strict diet are more in danger of becoming ill.
47. Whether it is true that making children study too young ruins their mind, stunts their growth, and makes them melancholic.
48. Whether it is true that there can be a fatal wound on the arm.
49. Why people say there is a remedy for everything except death.
50. Whether it is true that those who have hard nails and a tough hide live longer.

51. Concerning poultices for the wrist and neck dressings.
52. Why people say to those who drank a lot: "You must take some of the hair of the animal."
53. [There is no number 53; the error in numbering is corrected by putting number 56 twice.]
54. Why people say that during the first year of marriage one is in danger of getting scurfy, jealous, or cuckolded.
55. What calf fever means: why one shakes while drunk.
56. Whether it is true that one catches only once the plague, quartan fever, smallpox, measles, and scurf.
56. [See number 53 above] Whether it is true that clothes made with linen keep away lice and are not good for wounds and lesions.
57. Why those who rarely get sick become more seriously ill.
58. Whether it is true that one must not eat when angry, and whether one should eat very little when one is very hungry.
59. Why people say: "Light bread and heavy cheese."
60. Why people say: "Let him who cannot eat drink."
61. Concerning watered-down wine, and whether it should be given to people with a fever.
62. Why people say in Italian: *Qui [Chi] va pian, va san.*
63. Whether it is proper to say: "Bread one-day old, flour one-month old, and wine one-year old." Likewise: "Go to the fish market early and to the butcher shop late."
64. Why people say: "Beef when it's still bleeding, mutton when it's still bleating, pork when it's rotten; it's all worth nothing if it's not cooked."
65. Whether the first swallow of wine should be watered-down because it goes to the liver.
66. Against those who say that all bloodletting weakens the eyesight and those who say that moldy bread sharpens it.
67. If it is good to go through one's meal without drinking if one is not thirsty, and to eat a crust of dry bread in the morning to counteract the phlegm in the stomach.
68. Against those who say that in eating and shitting (speaking reverently) a man should hurry.
69. Whence it comes that heavy meat eaters have stinking breath.
70. How one should understand the saying that having the scurf is healthy; and whether it is better for apostemes to fester and burst than to be absorbed.
71. Whether one should do anything for smallpox, measles, and other children's diseases.
72. How it is that reading or writing immediately after meals hinders digestion and causes rheum.
73. [There is no number 73; the error goes uncorrected.]

74. Whether it is true that the frequent use of medicine makes people age faster; and whether it is bad to let children get accustomed to it.
75. Superstition of those who carry salt when they must cross some river or stream so that their wounds or lesions will not smart or open.
76. Why people say: "Uncooked chicken and raw veal fatten the churchyard."
77. Why oysters are appetizing, as are olives.
78. Whether it is bad to heat one's stomach after a meal, as they say, and to put a fur piece against it or feathers around it.
79. Whether a swallow of pure wine at the beginning of a dinner loosens the stomach.
80. Against those who maintain that one can heal a wound without seeing or touching the patient, provided one has the coat he was wearing when he was wounded, or by greasing the sword with which he was wounded, provided it is not rusting.
81. Against [Concerning?] those in whom the absorption and dissipation of apostemes (without coming to a head) is suspect; and if the pus has gone back into the body.
82. Whether by heating one's feet one is sooner relaxed; and whether one is sooner refreshed by drinking a little pure wine.
83. Whether blended wines are more inebriating.
84. Against those who think that redness in the face is always because of wine, and that water will not make it go away.
85. Why people say that a peach impedes, and the pit "unimpedes" [*La pesche ampeche, & le noyau desampesche*].
86. Whether it is right to say: "Butter in the morning is gold; at dinner, silver; and at supper, lead."
87. Flesh [meat] makes for flesh, poison for a din, pears for stones, and walnuts ruin one's voice [see number 56 of the 1578 edition].
88. Against those who do not allow one to change sick people's linens.
89. Whether lowering the bolster will hasten the death of the patient.

TO THE BROAD-MINDED AND STUDIOUS READER

Friendly reader, I had three principal reasons for publishing the index of the material I plan to discuss in my *Erreurs populaires*, although for the time being I am only bringing to light the first five books. One reason was to obligate myself to pursue such matters as I had promised. Another was that if someone was attracted to this subject and wanted to undertake a similar work, he would at least avoid the area I had staked out for myself and not (as the proverb goes) put his sickle in my harvest. For I can justly call it mine inasmuch as it was I who sowed these sayings. The third reason is to invite you, broad-minded and studious reader, to send me sayings similar to these, which I have been gathering for a long time from several people in several countries. Thus I hope to receive from people throughout the world who will read my index common sayings regarding medicine and health care, which they see are unknown to me because they are not in this collection. (For I have no interest whatever in errors concerning customs, management, government, and other affairs of human existence.) They can address them to me at Montpellier, if they do not have any further news from me, where I have the honor of presiding over the most famous medical university in the world. For which reason I was encouraged to labor toward the correction of the popular misconceptions that often cause trouble for young physicians and give them a hard time, especially since they do not yet have the authority to refute them on account of the small respect people have for their youth, even though they can be quite capable.

Such errors can be most harmful to man's health and even his life, and there are others that render physicians likely victims of calumny. Now, I am not saying that all the sayings contained in my index are erroneous. Several of them are proven and true. But ignorant laymen, not knowing the basis for what they say, are just as much in error; and in these cases I wish to examine their sayings in my discussion.

There are, therefore, among these popular sayings that I search out and gather, some which are totally false and erroneous, and others whose cause is unknown by the populace and are thus gathered under the name of *Errors*. This has been my intention, my guiding plan in the subject. Thus, I pray you, my friendly reader (whatever your station or profession), you who are neither opinionated nor obtuse, but broad-minded, refined, and studious, to be willing to help me. Favor me by sending whatever popular sayings you are able to encounter, and I will arrange them according to categories in order to discuss them, just as I do herein. I will then know that this labor of mine has been pleasant for you, and that you, too, wish to pursue it until I finish all that I have promised you. In this case, I shall abandon all other duties in order to give you this satisfaction, hoping you also will take from it great pleasure and profit. Farewell.

TO THE MOST GREAT, MOST EXCELLENT AND STUDIOUS PRINCESS MARGUERITE DE FRANCE, MOST ILLUSTRIOUS QUEEN OF NAVARRE, DAUGHTER, SISTER, AND SPOUSE OF KINGS, FROM LAURENT JOUBERT, HER MOST HUMBLE AND MOST AFFECTIONATE SERVANT, SALUTATIONS

Madame, there is a great dispute between the princes of philosophy, Plato and Aristotle, over the condition of the intellectual soul, which they admit openly to be celestial, divine, immortal, and separable from the body.[1] But Plato holds that it is of itself knowledgeable in all things and that the memory of them is erased and lost the instant the soul is submerged and mired in our moist and soft bodies. Then, as our bodies dry up little by little, the soul, once again clean and shining, remembers and recognizes all things bit by bit, as though learning them for the first time. For, according to Plato, what we call learning is but a remembering.

Aristotle, on the contrary, affirms that our soul comes to the body ignorant of everything, but capable and very quick to conceive all things, being a truly simple spirit, but in potency capable of grasping everything. He compares it to a virgin tablet on which nothing has been engraved, ready to receive all the colors and shapes desired. This latter notion has had more of a following than the former and is held as the true one among those who philosophize best. For if one became knowledgeable through the drying of the body, it would follow that there would be no need for doctrine, and that error would have no foothold in our souls (provided that the external senses were complete and in working order); but both these conclusions are patently absurd.

What need would one have of doctrine or teaching if the soul of itself became, or became once again, knowledgeable? And if it were only because of the superfluous moisture of the body that the soul did not know everything, whatever could be shown to it would not be understood or retained, and we would have to wait until it dried up to remember the things forgotten. In such a case, doctrine would be vain and totally useless (unless it served to put someone lost back onto the path), for after the drying of the body, the soul would still be as lost as ever, continuing in its forgetfulness. Furthermore, all people of the same age and makeup would have to be equally knowledgeable, since their bodies would be equally dried and their souls equally less moist. As for error, what place could it have if the soul knew everything, as long as the external senses did not deceive it by showing it one thing for another? The soul would not know what it had not yet discovered or recognized, but this is not error, for at least what it knew would be true, since all knowledge is verifiable.

Now nothing is more common and ordinary in the soul than error and false opinion. They must come from somewhere and make their way in from the outside, namely, from false doctrine and evil persuasion. It is true enough that the soul is able to forge error and lies for itself (as is the case in most men), abusing itself through ignorance. For in wishing to reason and argue about something, the soul, ignorant of some aspects and not sure of others, forms a faulty syllogism and then a bad conclusion, with which it remains satisfied, holding it in ignorance because it is unable to discern the true from the false. Thus is error engendered, and it will be more firmly rooted in the souls of the presumptuous—forgers of false opinions—than it will in the souls of the credulous, who accept false doctrine without argument or difficulty.

There, madame, is the source of error. This shows well that the soul is ignorant of itself and capable only of receiving what one wishes to paint or engrave upon it, whether good or evil, true or false. Just as water is without character, receiving all flavors indifferently, and just as white wool takes on all colors, so, too, is the soul modified with each quality. Happy indeed is the soul that meets up with good masters, especially for rudimentary learning, so that it will not be engraved, tainted, saturated, or impregnated with evil traits, colors, humors, or odors that are false, corrupted, and vicious from the start. For it is almost as difficult, if not impossible, to erase, repair, or reform false opinions inscribed and imprinted in a soft medium, which receives them very quickly, as it is to change the luster, tint, and color already imprinted in the countenance and complexion, or to alter humors engendered by pernicious nutriments. From all these come similar manners and similar actions, which, like foul odors, refuse emendation without first reforming the entire humor engendering them and thus offend not only the nose but also the mind of those having more sense.

Madame, I leave the instruction of the soul in the Christian religion for the time being to messieurs the theologians, who seek to engrave the faith deeply in it, to color it with piety, to bathe it in sound doctrine, and to perfume it with odors pleasing to God and profitable to its fellowmen, namely, a holy and exemplary life, in conformity to doctrine and proceeding from piety, having its strength firmly rooted in faith. I restrict myself to my vocation: caring for the human body in order to preserve it, to restore it to health when it fails, all with the help of the grace of the Almighty God, who has founded the practice of medicine and instituted the physician for man's needs. In which vocation I have for a long time (at least twenty-five years now) labored to two ends. One is to instruct youth in this science both by writings and by word of mouth, sincerely and diligently giving the rudiments, bathing my students in solid precepts, initiating them in the most secret remedies, exercising them in argument and clinical practice. The other is to extinguish and annihilate several false notions and errors (the offspring of ignorance) that have long had credence and vogue in medicine, surgery, and pharmaceutics—that is, among the professions of these three branches of our art. From such errors come much abuse and futility.

Yet these are small in number in comparison to the laymen's misconceptions about medicine and health, which are for the most part so gross, stupid, and wrong that they deserve laughter more than reprehension. Still, because there are some errors that are most prejudicial to human life, it seems to me that one must not scorn or hide them but must demonstrate to the ignorant layman how and why he is off the track and set him in the right direction. For he does not act with malice or with the intention to harm, but rather for the best (so it seems to him), even in embracing his error. It is the physician's duty to dissuade him from his false opinion and procedures, and to instruct him to do better whatever his duty might be, such as caring for the sick or standing by them faithfully, under the guidance and orders of learned physicians.

It is also true that what causes the evil will also produce the remedy. The evil (that is to say, the error engendered in the soul of the ignorant layman) doubtless came from hearing or seeing physicians say or do things that he now wishes to imitate, but without foundation. For, failing to note several and diverse requisite considerations and syllogizing incorrectly, he forges for himself false conclusions and errors which he holds as valid and drawn (as he thinks) from experience.

This is a very dangerous evil of which physicians are the cause because they have divulged their orders and prescriptions too freely to laymen, who have taken them down improperly and do not know how to use them correctly.[2] It is therefore the duty of physicians to remedy this situation, to which end I have labored considerably, instructing many ignorant laymen in these matters. But it has scarcely helped, since most are incapable of reasoning and argumentation. In the end I decided to show in writing the errors of those laymen who have strayed, and to take as judge one who will not be suspected in any way and who will be capable of judging and condemning such abuses. For if physicians judge what physicians condemn, it would always be the same old song. It is better that such a judge be someone who has good common sense but is not a physician, who has a quick mind and sound judgment, who has no interest in the dispute and no emotions that might cause him to judge otherwise than human reason dictates, and who has powers of understanding, expression, and discernment beyond those of ordinary laymen, in order to sound out and weigh the reasons I will be fully explaining.

Now, having for a long time pondered who this judge might be, madame, I thought that a princess would be most apt, because of the respect and authority she commands as well as for the greatness of mind and understanding widely recognized as accompanying the magnanimity of illustrious royal blood; because of this blood, princes and princesses in their superhuman actions, words, and reasoning manifest several traits of divinity that render them like minor gods, admirable and venerable among their inferiors. Also, our Immortal, Invisible, and Almighty God, who makes princes and princesses lieutenants on earth in order to make people observe their duties under the sword of justice and scepter of sound laws so they can fulfill such a charge, gives royalty a most generous heart and a mind more divine than human. Thus did it seem right to me that a person of royal blood would judge best those errors, especially a princess, who would take it upon herself more easily than a prince, who would be busy with other important matters. Add to that the fact that ignorant laymen, whose condemnation I am pursuing, will submit to it even more willingly, seeing that the difference between the parties will be better observed by a judge who is a sweet and humane princess, easily approachable and affable. And I have been able to observe, madame (after having well noted the excellence of your divine majesty), that you seemed to me to be the most fitting person in the world today, above all others because of the rare virtues each admires in you (the more than angelic wit, exquisite judgment, healthy curiosity, and studious desire to know all things), and because of the time you have to accept a recreation that will furnish you for a few hours each day much amusement in hearing and examining things I say against ignorant laymen to redress their errors. In this you will imitate, madame, as in all your other generous acts, the studious nature of your great aunt of the same name, the Marguerite of Marguerites,[3] most illustrious queen of Navarre, whom you are already surpassing in diversity and profundity of knowledge by your assiduous study, as is widely known. You will also imitate more faithfully your closer aunt of the same name, Marguerite de France, most excellent duchess of Savoy,[4] recently deceased, the most studious and knowledgeable princess of her age and time, to whom Your Majesty was sole heir of her love of letters. Thus, you carry on the most venerable name of the Marguerites of France, the most fecund nation in the world in great minds and persons of great knowledge. And all thanks to your grandfather, who was the father of arts and sciences, not only in his kingdom but throughout the universe.[5]

It is to the excellence of your judgment that I lay the charge. You who are purer than pearl (of which you carry the name), most formidable in capacity, vivacity, dexterity, and solidity. In speculation, research, and explanation of the most thorny and difficult questions, you are beyond your years and fair sex, as if a goddess. It is also to your beautiful soul that I lay the charge, so well instructed from infancy in all virtue and science, a soul pure and spotless, candid, sincere, splendid, adorned with the most beautiful qualities required by its grandeur and high station in order to judge errors that have in no wise tainted or sullied it. And it is the duty of the soul to judge errors. For one does not perceive blots because one has them; indeed, note that the crystalline humor, principal part of the eye, though seeing all the colors, has none in itself.

This is a function worthy of Your Majesty, more worthy beyond comparison than the judgment of Paris, Trojan prince, even though he had to judge between three goddesses. For the question of error is of much greater importance than that of the envious beauty of Juno, Minerva, and Venus.[6] So, too, the recompense will be more glorious, invaluable, and eternally praiseworthy. You will deserve to have an altar consecrated to your memory in the temple of health, where sacrifices of praise might be offered as long as the world exists, when you will have, on the basis of your judgment and authority, condemned these popular errors, thus rendering more secure the lives of men.

Madame, I submit all these qualities and procedures before Your

Majesty's eyes, entitling them *Errors*, for even though there are popular sayings that are sound and true, it is not commonly understood why they are so. Thus, in my work as a whole there are more errors redressed than other material. Now, it is customary for authors to list what is weightiest and most important, as your divine mind will doubtless be able to discern.

I most humbly supplicate Your Majesty to take well and accept with a sovereign countenance what I present Your Ladyship with great devotion for the common good, praying God that He guard and maintain Your Majesty in His holy blessings.

From the court of your spouse the king, and my most honored lord, this first day of the year 1578.

THE FIRST BOOK OF POPULAR ERRORS CONCERNING MEDICINE AND PHYSICIANS

CHAPTER I
EXCELLENCE OF THE ART OF MEDICINE AMONG ALL THE HUMAN ARTS; AGAINST THOSE WHO CONSIDER IT USELESS

We understand the human arts, both liberal and technical, to be those that man, inspired by God, has invented for his necessity, convenience, or recreation. Among these is medicine, the practice of natural philosophy on the human body, for which all the technical arts are invented, just as liberal arts are for the exercise of the mind.[1]

We except from all man's professions only the sacred science of theology, which we do not include in the ranking when we exalt medicine above all the human arts, for theology is not art, but science, and is not a human science, but purely divine, not invented by man but infused by God, concerning souls rather than bodies. It is eternal, infallible, immutable, having for object or subject Almighty God, creator of the world which He made from nothing for the sake of man. Of man we are considering the rational soul, the body, and the possessions given to him for the preservation of life.

Theology has as its principal aim the soul and, as its next object, moral philosophy. Jurisprudence, restricted to human law, governs the possessions and appurtenances of man, rendering to each his own. Between lies medicine, maintaining the body in health, driving out illness, and preserving from death (to the extent God permits).

If the excellence of a profession is determined by its object, as is proper, medicine holds the second place. For the soul is worth more than the body, and the body more than the clothes. I do not wish to argue here with Their Honors the magistrates, who indeed have power over the human body in matters of life and death, but their power is limited to withholding or imposing the death penalty, according to the crime. As for withholding the death penalty, if it is by pardon (as only the prince and sovereign magistrate can do), it is by the authority God gives him and not from the knowledge of law, as with the one who finds not guilty the man called into court and accused. This is not, properly speaking, saving or giving life, since an accused man found to be innocent does not deserve death. The power of sentencing to death is not praiseworthy, at least not compared to the power of saving life as a physician does (through the grace of God) in the case of many struck with grave illness who would probably die if not treated.

Whether or not this is feasible, and whether through the practice of medicine life can be prolonged, I will amply discuss in the next chapter. I wish here to show in passing the excellence of man in order to confirm the excellence of that art dedicated to his preservation. The principal dignity of man is that God has endowed and honored him with His own image and likeness, giving him an immortal soul capable of divinity, and that He has provided all things for his needs, convenience, and recreation, having made for us the heavens, the earth, and all that is in them.[2] For God has no need of anything He has made: all is for our use. Whence it is easy to understand that man is more worthy and excellent than the entire world. It is also true that the heavens and the earth, which had a beginning, will also have an end, aging like a garment.[3] Man will never end but will change his state from mortal to immortal a short time after the soul divorces itself from the body, taking it up again in a form more glorious than before and not subject to corruption.

Since, therefore, man is the most sacred thing in the world, the science devoted to his person is the highest, right after that concerned with his creator. Since man is the most sacred creature of all, the art or science that maintains him in life and health is the highest. This is a strong argument for the preeminence and value of medicine, based upon the worth of its subject. Now I wish to treat a few of its other qualities, similarly recommendable, such as its antiquity, necessity, and utility, along with the authority of those who have prized and revered it highly for these reasons.

As for its antiquity, there is no doubt it was there right from Adam's transgression, as soon as he had sinned and was subject to sickness. He was his own doctor, to whom God had given the knowledge of the nature of all things, having him name them according to their properties. Profane history attributes the invention of medicine to Apollo, the sun god, signifying that from him proceed the healing virtues of plants and other drugs the earth produces. This is why they say that Aesculapius (the first to make a profession of this art) was his son and the father of Machaon and Podalirius, vulnerary physicians (also called surgeons) during the Trojan War, which is among the oldest stories in the world.[4] Now, antiquity is one of the qualities that recommends things, provided it is unbroken; for if something were not useful or necessary it would soon come to an end. But one can see that until present times medicine has been well sustained, all the while increasing. It has been adorned and enriched through the industry of the greatest personages, not only philosophers, but also kings, princes, and other people of high station, as history proves by preserving the fruits of their labors. True enough, the Romans got along without physicians for six hundred years, having an aversion to them because of the cruelty of a few surgeons of Greece, a nation they held in suspicion.[5] But since that time physicians have been honored, respected, and well maintained in Rome, put among the ranks of nobles and knights.

As for the necessity of medicine, nothing could be more evident. Yet it seems that the excellence of the art has been questioned because it is neither expendable nor desirable in itself. According to moral philosophy, that which is desirable in itself (such as having children) is more esteemed than that which is esteemed for some other reason (such as having things for the children); so, too, because medicine (like the technical arts, without which we also cannot survive) is necessary, it seems less praiseworthy than music, for example, which is desirable in itself. With medicine, however, the more it is necessary, the more it should be desirable and the more its effects considered excellent.

Here is where its usefulness enters in, making it so commendable. Just as there is nothing more desirable than health nor more sought after than long life, medicine, seeing to the needs of both, is the most beneficial of any of the sciences to man's happiness. For, arguing to the contrary, an unhealthy man is useless to society, and he who lives briefly brings little profit. Now (as the Father of Eloquence argues),[6] we are not born for ourselves alone: our parents, relatives, and friends, our fatherland, even the whole universe, require of us some profit and usefulness.

It remains to confirm these arguments by the authority of those great men who have much esteemed and much exalted the art of medicine and those who profess it, giving it infinite credit by their writings. To this end I shall limit myself to the exhortation made in Ecclesiasticus, and to the warning of our good father Hippocrates, who should not be suspect in the matter for having been a physician, for he was never mercenary nor in anyone's service, but free and generous in the profession.[7] He first separated medicine from philosophy. For formerly no one was exclusively a physician; rather, the philosophers saw to illnesses and made remedies from the things of nature, mainly for their own use, as Celsus bears witness, because of their weakened bodies, worn down with continual thought and sleepless nights.[8]

Hippocrates was the first to separate this art from philosophy and to make of it a public profession, passed on by his successors, Diocles, Praxagoras, Chrysippus, Herophilus, and Erasistratus, who divided medicine into three branches to better accommodate the ill, distinguishing the mechanical or manual operation, called surgery, and the preparation of drugs, called pharmacy or apothecary, as they are still practiced today.[9]

But this was done, for the most part, by mercenary people whose names here cannot be mentioned in an argument commending the art of medicine, not even Galen's inasmuch as he was among the first to gain from it.[10] I shall limit myself to what the Grandfather of Medicine has written, after citing these words from Ecclesiasticus. They contain the wisdom of Jesus the Son of Sirach, who wrote in his thirty-eighth chapter: "Honor the physician for the need thou hast of him: for the most High hath created him. For all healing is from God: and he shall receive gifts of the king. The skill of the physician shall lift up his head: and in the sight of great men he shall be praised. The most high hath created medicines out of the earth: and a wise man will not abhor them. Was not bitter water made sweet with wood?[11] The virtue of these things is come to the knowledge of men: and the Most High hath given knowledge to men, that he may be honored in his wonders. By these he shall cure and shall allay their pains: and of these the apothecary shall make sweet confections and shall make up ointments of health. And of his works there shall be no end. For the peace of God is over all the face of the earth. My son, in thy sickness, neglect not thyself: but pray to the Lord and he shall heal thee. Turn away from sin and order thy hands aright: and cleanse thy heart from all offense. Give a sweet savor and a memorial of fine flour, and make a fat offering: and then give place to the physician. For the Lord has created him: and let him not depart from thee, for his works are necessary. For there is a time when thou must fall into their hands. And they shall beseech the Lord that he would prosper what they give for ease and remedy, for their conversation."

These divine words aptly conclude our argument for the dignity, worthiness, necessity, usefulness, and superiority of physicians and condemn those who have spoken ill of them and who scorn in them the greatness of God, who has wished to give man such relief. Let us now listen to what Hippocrates says on the matter.

This great man in his book *The Law* complains already that even in his own time medicine was less esteemed because of abuses. Look, I beg you, at how things are today. "The art of medicine," he says, "is the most eminent of all; but because of the ignorance of those who pass judgment on those who practice it, medicine has already been passed over in favor of all the other arts. The reason for this, it seems to me, proceeds mainly from the fact that there is no punishment meted out in cities for the malpractice of medicine, as in the other arts, other than dishonor. But dishonor does not sting the offenders enough; they are like actors in a tragedy and have the bearing, the countenance, and the dress of those they imitate. Thus, there are many physicians in name and reputation but few in practice. For it is necessary for him who truly must acquire the knowledge of a physician to satisfy these six conditions: talent, discipline, good behavior, sound basic education, love of difficulty, and the time required. . . . With these he will become a good physician, not only in name, but also in practice."

But ignorance is an evil treasure, an evil ring for those who wear it, and an empty daydream, etc. Pliny pursues closely this idea, taxing the commoners who do not distinguish between the good and bad physician, trusting those who babble the most, brag, and make a good show. "It is the case only in this art," he says, "that one believes immediately whoever says he is a doctor, even though there is in no lie greater danger. Yet one does not check into the matter, so pleasant for everybody is the sweetness of hoping for the best for oneself. Furthermore, there is no law punishing capital ignorance, nor is there an example of retribution in taking people's lives. They learn at our expense and kill people in doing their experiments, and the physician alone enjoys impunity in killing a man. What is worse, they blame the intemperance of the patient and happily condemn those who died."[12]

I thought I would bring up these arguments so it could be understood that it is not only today that some have but the mask and appearance of a physician and cause medicine to be less esteemed through their abuses, just as many other things in themselves good or neutral are decried because they are easily abused. Since I have demonstrated that through medicine one is able to prolong life, a most wonderful act, I wish to demonstrate more amply how this is possible.

CHAPTER II
WHETHER IT IS POSSIBLE TO PROLONG MAN'S LIFE THROUGH MEDICINE

This subject has always seemed most difficult. It has often troubled the greatest minds because it lies hidden and buried in the deepest recesses of nature, causing much difficulty for whoever takes it upon himself to delve into it. The arguments of those who treat it are so heated in one camp and the other that one scarcely knows what to believe. For, on the one hand, there are several arguments concluding that man's life cannot be lengthened, either by drugs or by the practice of medicine. On the other hand, physicians maintain that this is possible. In order better to resolve the doubt, therefore, I shall begin by upholding each side of the argument, and afterward, as an arbitrator, I shall pronounce my opinion.[1]

Among those claiming that man's life is limited, and that he cannot go beyond that limit by any means, we have first of all the most patient Job who, inflamed by the spirit of God, says this: "The days of man are brief, and the number of months is with you, Lord, who have set limits to the life of man, which he cannot trespass."[2] Aristotle also affirms as much in the second book of *Generation and Corruption*: " . . . the time and the life of each thing has its account set and determined; for in all things there is order, and all time and life are measured in periods."[3] And in the fourth book of *Generation of Animals* he says: "It is reasonable that there are periods and seasons, both for the gestation and for generations and life spans, which are counted in days, months, years, or other segments of time and are delimited by them."[4] When Averroës explains this point he says that all that is has of necessity a determined life span.[5] Since all the works of nature depend upon a certain order, without which they cannot exist in any way, and since art is by far inferior to nature in this matter (as Galen has so well argued in his book *Marasmus*), one can easily conclude that life cannot be prolonged by any artifice. Avicenna agrees with this in searching out the causes of our inevitable mortality, saying: "Death is natural for every creature, different in each case according to its basic nature, when it is no longer able to preserve its natural humidity. For each thing has its span delimited, different for each individual according to the diversity of its nature. And these are the natural terms. There are others that are curtailed, according to God's will, etc."[6]

If the end of life is fixed and assigned to each by God's command and ordinance, that is, by God's servant, nature, which is none other than the order established among the things of this world by His commandment, death cannot be circumvented by man, but only by the grace and will of Almighty God. Such was the case with the king Ezechias, whose death the prophet Elias foretold.[7] Because of his repentance, his life was prolonged fifteen years by the mercy of God, who promises in His law long life to the children who honor their mothers and fathers and are not ungrateful.

Now let us see whether, contrary to what we have deduced, the natural limits of life can be stretched and lengthened through the remedies and prescriptions of our art. For there are many reasons persuading us that not only the order of nature but also our own industry promote long life. First of all, the astrologers affirm this when they speak of elections, figures, and images. And this is confirmed by the experience, care, and diligence that physicians extend to several persons, who, by availing themselves of their remedies and a sound diet, preserve themselves in health and, though subject to infirmities, live a long life when they would otherwise have died young and never reached old age.

Plato and Aristotle (among the greatest and most serious of authors) relate that a man of letters named Herodicus, the sickest man of his day, nevertheless lived one hundred years through great artifice and a carefully chosen diet.[8] Galen also confesses his personal infirmity but says he corrected it so well that he was scarcely ever ill, at least after having limited himself exclusively to the practice of medicine, except that he was once or twice struck with ephemeral fever (that is to say, of one day) only because he overworked himself binding the wounds of his friends. And if we believe certain people who write of him, he lived one hundred forty years. It is no longer necessary to cite the authority of Plutarch, who points out that many weak and delicate people of the past managed to live a long time because of our art, and that several new cases are seen every day.[9]

One must not counterargue that several intemperate and dissolute people who have always scorned a sound diet have nevertheless, without any help from our art, managed to reach a ripe old age, for it is certain that if such people (wellborn and of good character) had lived by the rules and had availed themselves of our help in times of need, they would have grown old even later and stayed alive even longer. This is easy to prove from the fact that certain unhealthy people, either by nature or by chance, often live longer than the strong and healthy. Why? Because the strong, having too much confidence in their strength, live dissolutely, without rules and proper diet. The unhealthy are sober and continent, abstaining from harmful things and observing a sound manner of living under the physician's direction; this causes them to live longer. Whence the proverb saying that a broken pitcher lasts longer than a new one. On this Galen is right in saying that those who ignore or scorn a healthy manner of living will not live as long as normally ordained by nature.

For the science of medicine, purveying health and life, has such virtue that a man who rashly scorns its prescriptions will not only live in misery and all manner of sickness but will also cut years from his life and shorten the terms nature had fixed for him, hastening his death and (as they say) cutting his own throat. When such a man is on an unsound diet he either consumes his natural moisture sooner than he normally would, or he smothers and extinguishes his natural heat, upon which all things depend for life.

Now, if it is the law and the nature of contraries that they are bound to act upon an identical subject, and that if one is present, then the other must be also, it follows that if a person can shorten life he can also prolong it. Since it is well known that human life can be cut short by excesses, one may conclude that it can also be lengthened by sound diet and proper life-style. Even though the troubles springing from our constitution cannot be avoided, such as the emanation and continual dissipation of our substance (effected by natural heat), from which proceeds old age because of excessive and unavoidable exsiccation, this process can nonetheless be slowed down by our art, and we can stop the last day from coming too soon. Stop to think. Do we not see people on the verge of passing away who are kept alive a little longer by taking a small amount of malmsey, aquavit or imperiale, some concoction of alkermes, or some motherwort?[10]

Is not the period and quickly arriving last line of life deferred to another hour by such means? It is said of the laughing Democritus that, upon being requested by his servants not to have his house draped in mourning during the upcoming Thesmophoria feasts, he lengthened his life by the odor of honey or (as others say) of warm bread.[11] This prolonging of life through such small means is what our physicians point out, and it seems to be quite true.

We have treated the two sides with contrary citations and arguments. It is now necessary to quiet debate and to see what must be believed. In order to do this with the greatest of skill, it befits us to define life spans, some of which are supernatural, others natural, and still others upset (also called shortened or reduced). We call supernatural those life spans Almighty God has ordained and fixed purely of His free will, and these we are unable to duplicate by any art or advice, like the long life spans God ordained to men in the first age of the earth before the deluge for the multiplication of the human species, and especially to Noah for the restoration of man. Natural life spans are those given to people according to their diverse natures and their diverse fundamental constitutions, strong or weak, by virtue of which some live a long time and others a short time, according to the order of nature. They will attain these limits (through the grace of God) unless they live in a disorderly way or some misfortune befalls them. This constitutes the third kind of life span, which we call upset. It can come about at any age in strange and sudden cases, such as injuries, poisons, burns, falls, collapse, shipwrecks, plagues, and other common evils. Such misfortunes are for the most part unavoidable, and it is not up to the science of medicine to use precaution but to heal the suffering when it arises, if possible. For this reason, we leave such shortened life spans to the whims of fortune (which is nothing other, speaking piously, than the pure will of God, aside from the natural order, as we have stated elsewhere) and will speak only of the life spans we call natural, treating them more amply.[12]

All the philosophers and physicians agree that the length of our lifespan is measured in terms of natural heat and moisture, upon which all life depends. Now, in order to preserve such conditions within us, our good Mother Nature (as Galen speaks of her) has given us a marvelous power, which, through respiration, pulse, and the continuous supply of nourishment, stops the normal dissipation of our substance and natural moisture and maintains our body heat. But such power (which we call nutritive), being limited and not infinite, cannot always defend and preserve our moisture by substituting another. And so the body dries up little by little, and this moistening power becomes less effective and weakens day by day, until, in the end, the body ceases to take sufficient nourishment. Thus, as the members become more desiccated, the body loses weight and languishes; then, diminishing even more, it wrinkles, and this condition is called old age.

This is the basic natural law of corruption and death common to each body, for death comes when our primitive moisture, nutritive or natural, fails; and this is the end of life, which we call the natural end. As for our art, it is not one that exempts from death (Avicenna says), nor even one that is able to carry each person all the way to the most remote limit of his natural life span, perhaps one hundred or one hundred twenty years. But it does provide us with two things:

freedom from decay of the body (unless it is in an external form, such as the plague or poisoning) and the conservation of our natural moisture so that it lasts longer and is consumed more slowly.

These two things are within the power of our art, prolonging life until the natural limits are reached, according to the particular nature of each person. It does this in three ways. The first is to protect from outside heat, to stop obstructions, and to cast off excrements; by these means the onset of decay is prevented, or any decay which had begun is arrested. The second is to administer food and drink in proper substance, quality, quantity, time, and order. The third is to abstain from what, by consuming and taxing our natural moisture, quickly dissolves or dissipates our natural heat, such as excessive work, spices, staying up at night, worries, and diverse emotions, but above all, excessive carnal copulation, especially at improper times, and other similar things that one can and should avoid, according to the orders and prescriptions of medicine.[13]

But (you will say), that is not in doubt, for all will certainly agree, moved by the aforementioned arguments, that those who are temperate and take care of their health will live longer. This is no more than attaining without shortening it the end and term ordained by nature, although managing to do this is rare. What one is basically asking is whether the end and natural close of life can be put off through the art of medicine. I answer that life is not only preserved by our devices but also prolonged. It is reasonable for a contention to be more strongly affirmed and advanced when its foundation, principles, and efficient causes can be proposed, developed, and made even more solid. Now, if the principles of life (that is, the natural heat and primitive moisture) are unable to be reintegrated, they can at least be restored, repaired, and rendered more vigorous by our art, as the curing of hectics demonstrates, and by the amending of those dispositions preserving natural heat.

Thus, if by a moisturizing regimen (by freshwater baths and other similar remedies) the natural moisture that would otherwise be consumed more quickly can be preserved, and if natural heat can be contained so it uses more frugally its fuel (in the absence of which comes natural death), who will not admit that life, which would have been shorter according to nature, is prolonged by our art? I fully recognize and admit that the hard and spermatic parts cannot be substantially or deeply moistened, but it will be granted me that they can be moistened in their pores and empty spaces, into which nourishing moisture will penetrate; by such means the loss of natural moisture will be offset. This is much like lamps, when water is added to the oil so the flame will burn less voraciously.

Still, we shall prove most soundly that life spans can be lengthened. Of the body's dispositions or constitutions, the one that is most responsible for life is moisture, or rather moisture and warmth jointly, commonly called the sanguine. The opposite, commonly called the melancholic, has the shortest life. The conclusion is that if both types followed the same diet and personal care, the first would nevertheless live longer, since the term of its life span is greater by virtue of the principles of its engendering. Now, the virtue of our art is so strong that it is able to change little by little this natural cold and dry temperament into its opposite. This Galen teaches us in the last two books of *Preservation of Health*. Does it not follow directly from this that the length of a life span can also be preserved through the art of medicine? So much so that one who was unhappily born and condemned to a short life, upon changing natures, could have longer life.

From this one argument which everyone (in my opinion) understands, let us pass on to the others, namely, how one can expand the limits of all the times of our lives, resulting in the stretching of the entire life span. First, that the vigor or flower of youth can be greatly preserved through the art of medicine has been amply demonstrated by Galen. Furthermore, there are two main objectives in preserving health that are within our power: that of restoring lost substance with proper food and drink, and that of ridding the patient of the excrements that result from them. If one does not fail in either of these, the body will enjoy health and will be preserved for a long time in the strength of its vigor.

For the same reason, old age (completely inevitable for those who die a natural death) is lengthened through our art, so that one's decline and return to dust in extreme old age will come very late. From this we conclude that, as with all the periods of our lives (for the times of youth and adolescence can similarly and more easily be lengthened), so, too, with our whole life; we can lengthen our life span through medicine beyond what is ordained by nature. And these are the limits that God, principal author of medicine, has wished to make subject to this art, and which are in our power as long as God permits and does not cut short the thread of our lives. Just as in other instances, in contradiction to every law of nature set by Him, He sustains and continues life miraculously, without any medicinal aid, even without food and drink.

CHAPTER III
AGAINST THOSE WHO ARE OF THE OPINION THAT PHYSICIANS PROLONG ILLNESS AND DO NOTHING BUT DECEIVE PEOPLE

No arts are more subject to calumny than the military and the medical. These two are marvelously similar in other aspects, as we shall be able to see in several of the following arguments. For in order to explain more clearly the practice of medicine, I shall draw parallels from the waging of war.

In fact, even now it seems to me that I will be able to make use of one in the matter before us. It is that, if one has laid siege to a city and does not take it in the time decided upon, or as soon as those who are far from it think it could be taken (without having said so openly), even though the captain has done in the matter all that is expected, he will still be suspected or accused of negligence, cowardice, collusion, corruption, treason, ignorance, precipitation or sluggishness in his undertakings, improper conduct, pusillanimity, or some other flaw in his command, and all will be false. But those who so judge do not know the resistance the besieged can muster, the copiousness of the supplies they can have, the strength they can display, and the power they can have to defend themselves beyond what the besieger himself had thought (who, incidentally, could have been given false information, not only from spies and others who report on the state of the place, but from outside appearances giving the lie to what might be inside).

Similarly, when a physician besieges an illness in the body of a patient, he is often misled by outward signs and fair appearances so that, thinking he is at the end of his course of treatment, he must start all over again. For there are more evil humors and decay than he had foreseen, or the disease offers greater resistance than the physician had thought, redoubling its strength and repelling more and more every day his remedies and aids. So the illness goes on longer than had been predicted, and the patient does not get better as soon as the physician had promised, or as soon as those who have some knowledge in the matter had thought. It is, therefore, unfair to accuse him of ignorance, negligence, avarice, malice, or any other vice that might induce him to make the sickness last longer than it should.

Coming to the matter of ignorance, I trust that the physician one has sought is considered knowledgeable, expert, and good. For if he is not, one is wrong to call him in the first place and to put the life of the patient in his hands. Indeed, in such a case the patient could say, as Christ did to Pilate, "He who has delivered me to you has done greater evil than you."

As for negligence, I grant that there are good, competent, and wise physicians who pass quickly through their visits and the dressing of their patients, but I would never believe they do so to make the affliction last longer. I see it rather as an inadvertence that they could also have in their other affairs. And for this there are good remedies: solicit them more closely, encourage them to do their duty, beg them to come more frequently and to be more attentive, and give them a fellow helper that would allow them to be more careful.

The worst one might fear (in my opinion) is avarice, for laymen think that physicians regularly prolong illnesses, stringing patients along to get more money from them. Here I wish to stop a while longer to refute this false idea, the most erroneous of all. For in the first place I suppose the physician to be an upright man concerned with honor and reputation. I also expect him to want to profit in his profession, just as anybody would want to acquire goods honestly in his vocation. If he is an upright man, he would never wish to make his patients languish on purpose. If he is not, he should not have been called in to begin with, as I have already said above.

Suppose he is evil. Then he will have the objective of being sought after and being well esteemed for another objective: that of becoming rich. Now if he stretches out the illnesses he could quickly heal, he is not a clever man and works contrary to his intentions. But if he heals in less time than the others, he will be in greater demand and will have such a wealth of patients that he will not be able to treat them all, and people will rather give a crown to him than a teston to other physicians.[1] For who would not rather pay double, even triple or quadruple, and be cured sooner? If one gives ten crowns to the other physicians who manage to heal after a time, one would not begrudge fifty crowns to the physician who could shorten the time by half, a third, or a quarter.

But the truth of the matter is that it is not in the capacity of the physician to do as he pleases. He would love to have the power of healing by touching, or by seeing, or at the first receipt, or merely with a good diet or other simple means. He would have less trouble, would be more highly esteemed, and would earn infinitely more money. Good Lord, whoever had this ability would soon be rich!

Therefore, one must not think that physicians, driven by avarice, cause illnesses to be long. On the contrary, they gain more in gratefulness, reputation, and recompense if they heal the patients sooner. Stop to think! Are there not physicians who have relatives, associates, and good friends from whom they receive no pay? Do they heal them in any less time than the others from whom they do receive pay, since the illness is the same and the patient similar? He gains nothing from the length of such illnesses. It is enough that he does not lose the gratitude owed him for the good services he renders.

I will say more. When he, his wife, or his children are ill, it is entirely at his own expense. And do they never have long illnesses? Are they more quickly healed, all other things being equal? It is pure folly to think physicians foolish enough to prolong illnesses on purpose, even if they have but slight concern for their profit and honor. But it does happen often that physicians underestimate things, just as those who besiege a castle thinking they will take it in three days but who end up spending a month in front of it without slacking off or sparing any blows whatsoever. They might think a wall will not take ten cannon blasts, and yet it proves to absorb more than a hundred. They might think that the besieged only have food and munitions for a week, and yet they hold out for two months. What is known is conjecture based on appearances, examples, and observation, which can very often be misleading; but the captain in charge cannot be accused of failing to do his duty when he does all his art demands. So, too, with the physician, who is most excusable, especially when he misjudges the amount and the potency of his remedies. For it is this above all that makes an art conjectural, defined by Galen in several places as the mean between absolute knowledge and total ignorance. This is why one must interpret in a positive way and take in the right spirit the success of remedies that the learned, expert, diligent, and careful doctor orders in the most appropriate and the most accurate manner he is able, leaving the issue and outcome to God, who gives and takes away, strengthens or weakens the virtue of said remedies as He pleases, causing the illness to be sooner or later resolved, sometimes for the better, sometimes for the worse.

There remains malice, of which the physician might be suspected. But if there is the least occasion of rancor, hatred, or ill will between the physician and the patient, it is not wise to call in such a physician. For it is, on the contrary, necessary that the patient like the physician and that he be loved by him or that, if they have not previously made each other's acquaintance either by name or in person, they should in that case begin a close and deep friendship; otherwise the patient will not accept wholeheartedly the assurance of the physician, and the physician will not pay as much attention to the patient.

As for the willful malice of secretly harming a patient, if any physician is stained with this vice, he must be put in the same ranks as poisoners and obviously must not be called in on any case. But I think that people use this word "malice" in another sense here: namely, when physicians knowingly weaken and make patients who are on the verge of dying suffer unnecessarily by ordering abstinence and evacuation, not to help them but to display their power just for show; or when physicians hedge and protect themselves right from the start with the diagnosis that the patient is in mortal danger, while they are the ones that precipitated him into this state. These are (if I have properly understood them) the doubts laymen most often entertain.

Truly, it would be malicious, traitorous, and villainous if someone played this trick on a patient, neither more nor less so than if he threw somebody who could not swim into a river, putting his confidence in a rope thrown immediately thereafter to pull him out. For perhaps once submerged he will not be able to catch the rope or will not hold onto it securely or will not have the strength to pull himself out, and so the poor man will drown.

But it is not believable that physicians use such tricks, or true that they make their patients suffer with their remedies, which I presume are always properly imposed. It is the disease itself that continually weakens the forces of nature and increases its own up to a certain point, known as the vigor and sovereign state of the illness, after which, if the disease is a curable one, comes the decline or diminishing of the illness and of all its accidents, and the patient begins his climb toward convalescence, which we shall treat more amply (if it please God) in chapter seven of this book.

There are more modest people who do not say that physicians weaken their patients and endanger them, but that they make the illnesses last longer, either by being indulgent with them (that is, by being obsequious to their patients) or by making them dependent (through a long, trying, and severe illness). It is true that many patients would rather be healed slowly and treated gently; this excuses the physician, provided he protests to maintain his honor. As for prolonging the illness in order to elicit gratitude, this would be a fine piece of treason and wickedness. And so it is not believable that a physician who understands his position well would make an illness last. Why? Because the length of an illness cannot be gauged, and by entertaining its length, the physician might also encourage its gravity and end up in a much more serious predicament.

Ulcers, which are treated by the surgeon, are quite another matter, for he can maintain them without danger to the person; indeed, the internal parts of the body are better off for it, purging themselves through the ulcers, and there will not be any other illness than that of the ulcerated part.[2] Thus it is that we often order fistulas to be maintained and make openings or fontanels in several places on the body and keep them open for a very long time. But internal illnesses are another matter and must never be maintained if one is able to cure them; they must be dealt with immediately or as soon as possible.

The other point of calumny is that people are taken advantage of by physicians, that one could get better sooner and more completely without them, and that they are nothing but a hindrance. We have sufficiently refuted this notion in the first chapter with the authority of Ecclesiasticus; nevertheless, I shall add this metaphor (since I started comparing our art to that of the military) that there are forts that yield to the besieger when their supplies are cut off, others at the mere sight of a cannon, others at the first attack; but there are some which, after all this, still remain impregnable.

Now, if we were to argue that there are forts that surrender every day without force, we could say why bother besieging, assailing, fighting, bringing down walls? Why bother waging war on cities when we often see them yielding of themselves? It is an abuse and a stupid waste for a country, however seditious, to have soldiers, artillery, and other war supplies. It is but conceit and overreaching on the part of the people who live off that trade; we could well do without it. Yes, if all forts were weak and if there were no resistance from people endowed and supplied with courage and the other things necessary for their defense, they would surrender easily, like illnesses that do not need to be forced out with strong remedies but go away by themselves; and even the more serious ones, like burning fevers, when there is nothing in the body for them to feed upon and when the natural resistance fights bravely the tenaciousness of the illness. But with stubborn illnesses one must have help, employing a whole battery of remedies. And even with that support, one often makes no progress, and the sickness remains incurable. In such a case we must have no regrets. We must not say one would have recovered better without this or that, or that the patient was abused. For it truly would be an abuse if one were to promise a cure for an illness known to be incurable, one for which there is no remedy strong enough to vanquish it, just as it would be to undertake forcing a walled city with bare fists, or knocking down walls with a musket where a cannon or some similar instrument is needed and cannot be had.

These are the notorious abuses, a lot like the tricks to which simple folks fall victim, when empirical charlatans promise cures for any and every illness.[3] Such as these, it can be said, abuse people, but not reasonable, learned, expert, and honest physicians.

CHAPTER IV
THAT IT IS NOT A SIN OR AN EVIL DEED TO CALL A PHYSICIAN AND USE HIS REMEDIES WHEN ONE IS SICK

There is another kind of error, founded in pure superstition by some fools who think it is an offense against God if they call physicians to heal their ills, claiming that in doing so one is resisting and opposing the will of God, who is visiting them with an affliction for their own good. For by chastising the body, the soul is purged of its sins, and they say (as Guy de Chauliac tells us in an excellent chapter on the subject): "God has given it to me as it pleases Him, God will take it away when it shall please Him. Blessed be the name of God. Amen."[1] They submit their healing entirely to the intercession of the saints in heaven, making vows, giving alms, and saying prayers.[2]

This most erroneous opinion is easy to refute by what we have demonstrated in the first chapter based upon the book of Ecclesiasticus, where the sick are devoutly and wisely exhorted first to reconcile themselves with God, whom they have offended, then to seek help from a physician whom God has created and to whom He has given knowledge in order to be glorified in His marvels.

It is true that God sends us ills for our chastisement and has made us subject to them so that we might recognize our infirmity. From Him also proceeds healing through the means He has established in nature, giving to plants and other creatures the power to drive out and destroy diseases, and providing, on the one hand, agriculture for man's nourishment and, on the other, the science of medicine and the art of the apothecary for the purpose of preserving our weak and mortal lives.

So these are things we must not scorn, and which the prudent man will not despise. To do otherwise is to tempt God and to wish foolishly that He perform miracles at our bidding. For he who says, "If God wants me to recover from this illness, I'll get over it without medicine, and if I am to die from it, the physician is not going to help me," might as well say, "If I am to live another month and it is thus ordained by God, I will live without food and drink, so it is not necessary to take the trouble; for if I am to live so long, it is impossible that I die, even though I eat nothing." This is foolishness and great temerity, to presume that God will perform a miracle, or even to try such a thing when one has on hand food ordained by God for the nourishment of the body. Is it not tempting God to expect to see what He will do against the order of nature? He will let him die of hunger along with his folly, and the poor fool will feel the effects of his ill-conceived, stupid, and crazed notion that God would maintain him in health without food and drink. Of course, if God wished, He could do it; but we know that ordinarily He wishes us to use food, and that is what we should do and not expect unknown and untested extraordinary means to save us because of our foolish desires.

Thus it is with medicine, ordained by the Almighty for the healing of the sick and the preservation of health. For whoever wishes to be healed otherwise and is of the opinion that if he is to get better he can do it without a physician, even though he has the wherewithal, is tempting God and expecting Him to perform a miracle, while scorning the natural means He has ordained for the curing of illness. It is as if his house were on fire and he did not want any water thrown on it, saying "If God wants it to be saved, the fire will go out some other way."

CHAPTER V
ON THE INGRATITUDE OF PATIENTS TOWARD PHYSICIANS

Ingratitude is most odious to God and to man. In fact, it is considered such a horrid vice that he who is ungrateful is afflicted with all the world's evils. Now, this vice is so commonly attributed to physicians by laymen that I am often astonished that there are any people left with hearts generous enough to wish to be physicians, since it is the profession most subject to calumny (first cousin to ingratitude).

But we know friends and reasonable, honest, and grateful people who make up for this unfortunate state of things and keep us of a mind to practice the profession in spite of the fact that several others are horribly ungrateful to us. For we find people so courteous that they will profess publicly and often that they owe their lives (after God) to such and such a physician and, having recognized the industriousness and toil of the physician to be worthy of his fees, will nevertheless confess openly that he could not be recompensed even with everything they have, as is truly the case. For if they owe their lives to the help of a physician, and if life is worth more than all their possessions, they could not acquit themselves of this debt if they were to give to him all they have. But the mainstay of the recompense is the gratitude they show to the physician, saying how much they are obliged to him and owe him their lives. Indeed, it is just as if the sword were taken from the hands of one who was ready to kill them or the cord from one trying to strangle them. Would they not owe him their lives? Would not all their possessions go toward rewarding him?

And then you hear people say: "I paid my physician well, even overpaid him, having given him so much a day; I owe him nothing if he has cared for me well, since I rewarded him well." Ha, poor man! What is given to the physician is a small token of the wealth and aid received, for to pay for or reward the fruit of his labor if he saved you from death (as he is able to do, by the grace of God) is not in your power unless you give your life for him, even though he did not expose his own to save you from death. Thus, you remain ever in his debt, and you must with gratitude acknowledge it, admitting your obligation.

There are some who will find it a little hard to take when I say the physician saves lives and keeps us from death, even though this is quite obvious. For let us suppose that an injured person is losing his blood in great amounts and that he will probably die from it if it is not stopped. Will not the person who puts his finger in the wound and holds back the blood be saving his life? All the more so he who holds it back with medicine and in the end closes the wound, which of itself would not heal. And he who clears intestinal obstruction or stops vomiting or some other pernicious and serious discharge, or he who bleeds a person suffering from pleurisy or saves another stifling from quinsy.[1] Take him who pulls from a fire a child who has fallen into it and would surely be burned alive if he were not rescued! One must esteem no less physicians, who see to the internal ills and help nature secretly with diverse methods of which the effectiveness is seen only through the results. These are (as Herophilus used to say) the hands of God.[2] For He raises us up and withdraws us from death by means of the remedies the physician calls to his aid. Is this not a feat more divine than human, and one that cannot be sufficiently recompensed? This is why Ecclesiasticus is right to proclaim: "The skill of the physician shall lift up his head: and in the sight of great men he shall be praised," and "He shall receive gifts of the king."[3]

The principal acknowledgments patients owe him are honor and gratitude because of a great obligation, not a begrudging admission that he is sufficiently rewarded with some money. But some patients act worse: after having been healed through the care of good and loyal assistance, they cannot endure hearing people say they are indebted to the physician, and they practically hate the man who has saved their lives. Oh extreme ingratitude!

But this is not only the case today. Hippocrates in his letter to

Damagetes has Democritus speak thus: "I think, Hippocrates," he says, "that several aspects of our science are subject to calumny and ingratitude because the patients, if they escape their illnesses, attribute their healing to the gods, or to fortune, or to their strong constitutions, robbing all the honor from the physician, whom they often hate afterward because they are angry and indignant over being considered in his debt. And besides not wanting to attest to or to admit their obligations, they are happy that those ignorant of the art (who nevertheless exercise the profession) are by the same token stung with envy, and so forth."[4]

This applies perfectly to our own times, for most patients attribute their healing entirely to a saint in heaven to whom they made a vow, and yet often enough their wishes are not granted, proving what the Italian proverb says: *Passato lo malo, poi è gabbato lo santo.*[5] Likewise, they make grand promises to the physician during the height of the illness, promising mountains and marvels: everything will be done with gold and precious stones, and he will have a handsome yearly stipend. In short, he is promised numerous rewards. But as soon as the patient recovers, he begins to think that the doctor had no part in it, that he could have gotten better without him, that it was the vow he made that caused the healing, or the good care of the attendants or the apothecary (who will want to claim all the success), or the good soups, or his good and solid constitution, or a chance happening, such as the excesses he may have allowed himself and to which he will wildly attribute his recovery.

In short, the physician will have the least part (if any at all) of the honor, gratitude, and reward. For as concerns the promises, once the man is better he is going to think how much the sickness has cost him, that he has spent such and such a sum, and that there will be so much interest. Thus, he forgets his obligation to the physician, to whom he even charges part of the cost, estimating it superfluous, and holds a grudge against him for having kept him in bed, taking so long to perform his task when he could have acted more quickly and for less money. So much so that, by the patient's reckoning, the physician is indebted to him, and if he could find judges on his side who could carry enough weight, he would have him pay the expenses.

Is this a proper reward for services rendered? Has anyone seen such ingratitude? No, unless it be that of a man who, hanging himself out of despair or for some other reason, takes to court for damages to the cord the person who came to his aid and cut him loose. Or that of a drowning person who turns to one who tore his clothes while saving him only to demand that they be repaired. Thus, those who are our debtors are billing us and have neither gratitude nor graciousness for our having served them well, preferring to say that an ignorant valet or chambermaid is more the cause of their getting better than the sound care and diligence of the physician. And this attitude is attributable to two causes: either they are so dull they have not the capacity to understand, or knowing it well, they are ashamed of not having the will to recognize and admit it. Whatever be the case, it is a most odious ungratefulness both to God and to men.

CHAPTER VI
THAT LAYMEN HAVE LITTLE RESPECT FOR A PHYSICIAN WHO DOES NOT TREAT ACCORDING TO THEIR DIAGNOSIS; THAT THE LAST-USED REMEDIES GATHER ALL THE GLORY; AND HAPPY THE PHYSICIAN WHO ARRIVES AS THE ILLNESS IS WEAKENING

This error is closely related to the preceding and is even the cause of the ingratitude we mentioned above. For if one does not heal according to the ideas of the patient or of those who visit him, it is as good as nothing, and no gratitude is shown.

Now healing, according to popular opinion, falls into one of two types. The first is to heal in less time and before it was expected, as in the case of an illness that normally lasts so many days but is healed much sooner. For otherwise people say: "The illness has run its course, the physician was for naught in it, so he recovered with time."

Poor people! Do you not see that in any given type of illness some are short and others are long? There are tertian agues—and continual ones also—that last one or two months.[1] You suppose that the tertian fevers should not last beyond their stretch of fourteen days, and continual fevers seven, eleven, or fourteen, as you have heard physicians say, which is the length of exquisite fevers.[2] But you do not know that out of a thousand there are not two alike, and most are confused and mixed, which causes them to be much longer, as are all illnesses springing from a mixture of diverse humors.[3] Believe me (and this is true), if tertian ague finishes three weeks or a month after having been weakened with our remedies, it could possibly have lasted two or three months without them (as we see with many others).

Is this not proper and fitting treatment for the patient? No, for to hear him talk, we have done nothing if we have not done more than he had expected.[4] For he thinks that the physician is able to do with the illness as one does with stirrup leather, which one stretches and shortens at will. Is it not enough to shorten it by a fourth, a third, or a half, and arrest or soften the diverse complications that arrive in all sorts of illnesses, and make it so that one manages to have the upper hand, to ensure the best outcome possible, and to overcome the illness, whatever the price?

Here we fall into the other extreme of popular opinion, which holds us without worth if we do not heal those thought to be at death's door. For even though the illness be fatal, as is every illness we call acute (that is, cutting, that goes quickly and has terrible complications), if the patient or his visitors are of the opinion that he can be cured, and he in fact is, it is as nothing; but if, on the contrary, the patient dies from it, it is the fault of the physician. For those present were convinced (even though the physician said the opposite in his prognosis) that he could be cured. But if it is thought that he must die, or if he is already thought as good as dead, the physician has sport enough, for even if he only orders soup with a few little confections, especially restoratives and heart strengtheners (although this might not be apropos), he has done a masterpiece. Here is a beautiful piece of healing! He has cured so-and-so, whom all thought marked for death! He has resurrected him! He is a great personage!

But here is the sad part. This same doctor may have at the same time another patient who is not thought to be so sick, because his illness is more hidden. He goes beyond the call of duty to make him well and, in order to succeed, uses all his resources to save the patient, who, he knows, is in greater danger than is thought. In the end he dies, contrary to the expectation of laymen. Here is your physician suddenly losing his reputation, and people say: "He did too much in this case; the other one was better looked after." Thus, we never do anything right if we cure contrary to popular hopes and expectations.

The other error proposed in this chapter is attributing to the last-used remedies the entire success of the cure, just as people accuse the occasion of the illness to the last thing they did, as when they have eaten some fruit, salad, or other less ordinary food and immediately afterward become sick. Even if it is an illness that has been going on for more than a month, they claim that that one thing is the cause of it, without bothering to consider an indefinite number of other preceding disorders, which have all had their part.

For bad humors accumulate little by little, until they attain a certain mass which nature can no longer reject. Just as a glass gets filled up with several drops of water, which it will contain up to the rim, but, once full, will begin to let the water overflow one drop at a time, so, too, the least little addition to what nature had still been absorbing makes it succumb, like a mule under a burden, when but a fraction is added to what it usually carries.

Thus, it is not the last straw, or disorder, that is the cause of everything; that which preceded played its part. Not any less than when cutting down a tree it seems in vain that a hundred strokes of the axe are delivered, but the hundred and first makes it fall. If one were to say that this sole stroke felled the tree, would one not be doing injustice to the others? When a tower has absorbed a thousand cannon blasts and at the last blast finally crumbles, has the last blast done more than the first?

It is just the same when one judges the remedies that blast the illness and drive the disease from the body: the last, whatever it might be, has the honor and praise of the stupid populace, who speak thus: "They bled him, purged him, clysterized him, drugged him with a thousand things inside and out, all for nothing. Finally they gave him such and such, and he was cured." Poor fools! If all that had been done at the start, it would have accomplished nothing, but after so many other remedies that weakened, shook, and uprooted the illness, the least thing in the world would have made it go away.

Just as with people who are besieged and who cannot take any more, if one kills but one more of them, they surrender immediately. And then it will be said that all the battering, the attacks, the cutting off of supplies, and other good methods of vanquishing them served to no avail; he alone who fired the last shot, and who yet only killed one of the least soldiers, did it all. If he had killed the chief this would have been something else.

Thus, a note pinned on the collar, or remedies placed on the wrist, will have the honor of having cured the fever of a patient not able to be cured by scores of diets, prescriptions, and other applications.[5] It is that the illness was hanging on by a thread, which was able to be broken by the conviction and belief that the patient had in this last remedy. But if it had been applied right from the start, the patient would not have been cured, even if he had had a hundred thousand times the belief in it, plus a strong imagination. For imagination can play a large role in getting better, but it is not everything and can do nothing alone.[6]

This is how people rob the honor due certain true remedies: by misjudging their success. Because one wants to be cured as soon as some application is made, anything not working is thought to be useless. The sole author of the cure is that which immediately precedes it.

And yet it is commonly said (which is the third point of this chapter): "Happy the physician who arrives as the illness is weakening." For whatever he does, since the cure awaits him in the doorway, its happening is attributed to him. And even though the physician did not do anything or prescribe anything, still it will be said that he is the cause of this happy turn, and that if he had come at the beginning the patient would have been cured just as quickly.

But if the physician is prudent and modest, he will not bathe in this honor nor consent to the larceny and detraction done to those who have properly treated the patient and who are the true authors of the healing, but he will point out to those present that the past symptoms were characteristic of the illness, that they had run their course, and that because of the good care, all is cleared up for the good of the patient. If he does otherwise and wishes to attribute to himself honor or accept it from ignorant laymen, he is doing himself a great wrong, and he is the more to blame. For whatever importance and reputation he may have, it could happen that someone will call in another doctor toward the end of an illness he has been treating, and that this doctor will play the same trick on him.[7]

Thus, let all take note and content themselves justly with the honor due them without robbing their colleagues, or symmysts (that is, professional companions), rendering sound and fair witness to the praiseworthy actions of each, and considering themselves nonetheless fortunate (when they arrive as the illness is waning) not only in being spared much toil, but also in being able to take part in expressing gratitude to all those who labored to its end.

CHAPTER VII
THAT THE PROFESSION OF THE PHYSICIAN IS SEEN IN AN INAUSPICIOUS LIGHT WHEN SOMEONE DIES OF AN ILLNESS THAT OTHERS ARE CURED OF[1]

Every day Terence's proverb is found to be most true: there is nothing more foul and unjust than an ignorant and untrained man.[2] And this complaint is confirmed in our art more than in any other, as can easily be seen by the almost endless list of errors constituting the matter of this work.

The error I am about to deal with now is very common and is linked to the preceding one: that the ignorant attach no value to a cure unless the patient improves against all hope. For if someone dies of blood loss, diarrhea, dysentery, of a tertian or some other intermittent or continual fever, or of pleurisy, the fact that others have been cured of them causes people to think the death is the physician's fault, either through his ignorance or his negligence.[3] This is why we hear a lot of murmuring when someone dies of a gunshot or some other wound in the arm or leg, because people consider fatal only those in the head and in the trunk, that is, the chest and lower abdomen.[4]

Thus, if the physician heals these serious wounds, laymen will infinitely esteem the procedures and industry of the healer, just as they will decry these same procedures if, on the contrary, a patient dies of wounds to the arms or legs. And there is constant fear that the wounded patient will die from hemorrhage (that is, a flow of blood), gangrene and sphacelation (called Saint Anthony's fire), or other complications. As if all afflictions similar in kind shared the same peculiarities, and as if there were not an infinite number of differences among illnesses, just as there is among the human species and among all other things.[5] For man is but a type of animal, as a wound is but a type of affliction; and as there is infinite diversity among men, so, too, among wounds, wherever they occur. This I say purposely so that it will not be thought that the location alone makes any difference, even though the location, through its diversity, makes the complications vary infinitely. And even if one were to grant all other things equal, both the location and the wound, there are still a thousand circumstances springing from peculiarities of the subject (the wounded person) because of his constitution, corpulence, age, strength or resistance, habits, previous or present manner of living, including his normal dwelling place, food and drink, amount of sleep, work and leisure, meals and voidings, as well as his coitus, along with his emotions and his dealings with others. To which must be added the conditions of his humors, which are the main cause of the good and ill that befalls wounded people. For people with bad digestion always have a worse time of it.

Thus, if two people identical in every way cannot be found, or even one who is like what he used to be at a younger age and in another season of his life (seeing that we change from hour to hour), how can one argue for the similar, which is never found except in the species and then, as they say, only roughly and never exactly the same in each individual?[6] It does not take much to throw off an equilibrium, as a half of a grain will spoil a crown.[7] This is why one patient, unfairly compared to another who died from an illness said to be similar, might well be at the point of dying but, because of a condition better only by a hair, will tip the scales toward recovery and, once on that road, not stop until he goes all the way. Another patient, however, because of a hair too much of a bad condition (a strange thing, and hard to understand) will slide into death's grip and not escape, whatever help one might give him, because he could not recover from that single blow.

This is why some die of pleurisy and others recover; and the same person one time will recover, even when it seemed more vehement, and another time will die of what seems a lighter case or actually is less violent. So, too, with wounds to the head, chest, and stomach: some people will recover who another time would die even of arm or leg wounds, which are considered less serious. It is wrong to relieve our regret by saying, "If so-and-so had dressed the wound, the patient would not have died from it, for we have seen him recover from worse," because it may well have been the same surgeon or physician who treated him in both instances.

Stop to think! There will always be the rejoinder from the wild, opinionated, and excessive malcontent that the physician or surgeon could have saved someone from dying if he had been more diligent and attentive, or if he had watched more closely and more carefully (as in the past, when he used to be more helpful and dutiful, took more time, and had other such qualities befitting a physician who treats his patients dutifully).

Now, I do not deny that such instances of poor care cannot occur, for in fact they are often the cause of these different outcomes: that one recovers while the other dies.[8] However, they also come about for the most part because of the patient's condition and because of a particularity that is completely hidden. This is why the one who is there to dress the wound gets all the glory—an unjust and unfair state of things!

So, as I said right at the beginning, there is nothing more foul and unreasonable than the judgment of ignorant and unskilled people.

CHAPTER VIII
AGAINST THOSE WHO SCORN PHYSICIANS FOR DIAGNOSING ILLNESSES INCORRECTLY; AGAINST THOSE WHO WISH DEATH UPON PHYSICIANS WHO DIAGNOSE THEIR ILLNESS AS FATAL; AND WHETHER IT IS PROPER FOR A PHYSICIAN TO ABANDON THE PATIENT HE JUDGES CERTAIN TO DIE[1]

It is for God alone to know and foresee the future with certainty; and it seems that all the rest has been freely communicated to man, whose mind is capable of grasping everything save certainty about the future. To God all things are present. It is quite true that, through the observation of natural things, which often run to a similar end, one is able to guess roughly what will happen. Also, prudent and very careful people foresee the evil or good that can result from a given enterprise. Thus, farmers predict a good or bad season, and the sailor foresees good or bad weather. But inasmuch as nothing is certain, given the inconstancy and frequent changes in corruptible things, either through chance or through our own fault, through the unknown secrets of nature or the providence of God (who, in a moment's time, changes and overturns the normal outcome of things), it is not possible for men to predict the future other than by conjecture and fallible arguments.[2]

As for illnesses, we can sometimes predict death with much certainty. Health, however, cannot be assured, because a curable illness soon becomes incurable, either through the fault of the patient or of those attending him who do not carry out completely the sound advice of physicians. For this reason several short and curable illnesses turn into long and fatal ones. This is why careful physicians, in order to avoid calumny and reproach from the commoners, and to avoid misstating the nature of the illness (and thereby protect the rules of our art from suspicion and condemnation as false), predict life or death according to the strength they find in the patient, with the condition that some other complication does not befall him.

Also, we are unable to say that the physician will affirm with certitude that an illness is curable or not, and that he will use good and appropriate remedies.[3] Yet God (who is above all) will allow a person to get better who, by the laws of nature, ought to die, and vice versa, for His judgments are inscrutable. We must listen carefully when Galen points out that we should not prescribe anything for those cases we consider hopeless lest the remedies and the art be scorned or defamed. Yet it would be a great inhumanity, unworthy of a physician (who should be most helpful, full of piety and compassion) not to visit those who are, in his judgment, bound to die, abandoning them on a simple prognosis. In fact, it seems to me that the patients in this state have the greatest need of visits in order to be encouraged to bear the illness they must endure. Several are in danger of death for not obeying their physicians or through the fault of those attending to them. I say nothing of the secret and hidden constitutions of some patients, which are not easy to understand and, if not understood precisely, hinder healing.

For this reason it is most necessary for attendants not to forget anything physicians order, even things that seem of little importance. Nothing must be added to their directives nor anything suppressed; the attendants must observe them wholly and diligently, failing in nothing that has been ordered. If physicians were left to see to their patients, and if there were not so many people making medicine their business, such as midwives, attendants, apothecaries, barbers, and an endless number of ignorant laymen, there would not be so many illnesses, and our prognoses would be more accurate.

But I am more amazed by the patients themselves, who prefer (for the most part) to take advice from idiots rather than from well-known physicians. I admit that several physicians do not do their duty as they should: there are those who, negligent and without compassion, try only to fill their purses. They trot from house to house, concerned less for another's loss than their own profits, with no modesty whatsoever, visiting apothecaries to drum up business, pretending and bragging that they can recognize every illness by examining the patient's urine. Through flattery and deception they slyly trick patients and almost anybody who wants to be deceived and abused.

This is the fault of the magistrate, who does not chastise misdemeanors committed in medicine, giving as free a rein to ignorant and dishonest charlatans and impostors as to learned and well-bred people.[4] Given the faults and mistakes of patients and attendants, one should not be surprised if the most experienced physicians fail sometimes in their predictions. To this confusion must be added the diverse constitutions of the patients, as we have said above. On this matter Celsus has pronounced correctly: "Medicine is founded in conjecture, and the reason for the conjecture is that although a condition is seen frequently, it can still surprise us. Thus, if a condition nearly always presents itself in a thousand patients, little account is taken of it, for in such a large number of people it will appear in different ways. What I am saying must be applied to curable as well as fatal illnesses. For hopes are sometimes false, and a patient who the physician was sure would live dies. Sometimes remedies that usually promote healing will make matters turn for the worse. It is not possible for man to avoid these incidents (given his limitations) in a complexity as great as that of the body. So, to the art of medicine it is necessary to add faith, which is always more helpful to a patient than doubt, and to accept what Hippocrates says: that diagnoses made of great and sudden illnesses (called acute) are more error-prone and uncertain than those of long and less violent ones."[5]

Once in the past, in consultation with another physician on the illness of a great lord and marshal of France, I predicted right from the beginning that his illness would be long or fatal.[6] I was calumniated because of this, especially since (thanks be to God) he did not die but recovered soon thereafter. Still, I thought I had accomplished, along with those who were helping me, a masterpiece: we had converted a long fatal illness into a curable and short one. Laymen cannot understand this, nor do they consider such a feat for its true worth. For if you say that the illness is fatal and death does not come, they say that you misjudged the matter.

But stop to think! Is not the plague a fatal illness? Yet several get over it. So, too, the continual fever, pleurisy, a wound through the body, and several other diseases properly considered fatal (it is not that all inevitably die from them, but they do for the most part). Is it not the height of insolence to blame a physician for a diagnosis of pestilential fever, with a prognosis of the purples,[7] followed by frenzy, convulsions, and lethargy (as in the case of the lord mentioned above), as fatal or long? On the contrary, the physician should be infinitely praised for having, through his diligence, observation, and sound remedies, converted the fatal illness into a curable one, the long into a short, turning away complications with such dexterity that there was but a shadow of a light affliction of frenzy, convulsions, and lethargy, even though such complications are fatal in themselves, as the most brilliant and experienced physicians well know.[8]

I now come to the other topic: those who wish death upon the physician who diagnosed their illness as fatal. It is as if he were a judge who in the past sentenced them to death, but they escaped punishment and now wish death upon him and would gladly have him hanged if they could.

But the cases are very dissimilar, because a judge sentences people to death and plans to follow through on his sentencing, while the other only believes that the illness will cause the patient to die. It is like a person who is watching two men fighting or gambling and predicts which one will be beaten. Can his prediction be said to cause the one or the other to lose? Someone who escapes from the sentence of a judge through grace and pardon granted by the prince is like one who escapes a fatal illness by the grace of God, through the help of the physician. If one claims that the physician was in it for naught, that the patient owes him no thanks, and that his hour had not yet come, I in turn will say that the person pardoned owes the king no thanks either, since events proved that the condemned man could not have died either, for his hour had not yet come.

It is better to speak in other terms, and to compare the physician not to a judge but to a king. For just as God, not wishing yet to take from the world this criminal sentenced to death, puts in the heart of the prince a desire to save him and to grant him pardon, so, too, in the case of the patient afflicted with a fatal illness whom He does not yet wish to call, God places in the hands of the physician the wherewithal to cure him and blesses his remedies.

Thus must one always show gratitude to the physician for having predicted what seemed to him to be dangerous, because it is always better to be wary of death than to trust in it. There are several reasons for this wariness, but one of them concerns the bequeathing or inheriting of possessions, estates, and nobility, none of which should be exposed to chance. Another reason is the role of the physician, which is to be highly diligent and perspicacious in attending to his patient and in foreseeing and pointing out dangers.

Still, there are some who are quite careless in these circumstances, and who become negligent because they feel absolved from any criticism once they have diagnosed fatality. This is a big mistake. Celsus points out a much better path. He says: "When the danger is great, yet still without total despair, it is necessary to inform the patient's relatives, so that if the treatment fails it does not look as though one was unaware or negligent of the fatal aspect. But both the prudent man and the braggadocio benefit when they build up a little thing, because they seem to have done much more. And it is a good idea to identify and treat the quick and easy case lest the illness, which is in itself small, become more serious through negligence on the part of the one treating it. This is how one must proceed so as not to incur hatred upon the physician who has wisely warned those concerned of the danger he considers the patient to be in."[9]

As for the third point proposed in the title to this chapter, I am still of the opinion that one must never abandon a patient because of some complication that develops: one must stay with him until the very end. And just as you must not fail to give him nourishment regularly as long as there is life in the body (even though you know very well he will die in the next hour), so, too, you must always administer some little remedy, without in any way annoying the patient with matters of importance. For several recover against all human hope, who, had they been abandoned, would certainly have died (speaking from a human point of view), as do those who are buried alive because they are thought to be dead.

Galen and Celsus do not contradict my statement when they warn us (as we have said above) not to touch those we have no hope of curing, for fear that the remedies which have profited many will be attacked. For they understand by this the remedies that are extreme and subject to calumny, such as bloodletting, purgation, lancing, cauterization, and so on, not the small and gentle ones, which must always be continued until the end, be it happy or sad. The reason? Nature most often has within herself some hidden and latent virtue that shows itself after withstanding a thousand assaults, and to which she gives the upper hand, as in the case of a house or an entire city blazing up from a single spark. Only a wisp of strength must hold firm and be supported properly in order to drive illness out little by little and to restore health in its full mastery.

For this reason we must never abandon the patient, both for the reasons stated above and for unexpected recoveries witnessed daily in many people, to the great embarrassment of those who had given up on them and left them for dead, affirming obstinately (without any exceptions or qualifications) that it was impossible to save them. This is why an ignorant person, or a less knowledgeable one, even though he does nothing worthwhile, so long as he does not abandon the patient, can carry away with him the glory of having saved him.[10]

And so one must be prudent in one's prognosis and point out the nature of the illness as one that few people escape. But one must also point out that the force of nature is incomprehensible, and that God, who is above all things, often does miracles. Sometimes a physician will give up on patients he thinks are incurable and bound to die, often because relatives or attendants are also convinced of the hopelessness of the case (even more firmly than the physician), caring little whether he continues his visits so they might avoid the expense. This is particularly true when he is expressly commissioned on a daily basis, in which case if he wishes to leave they do not entreat him to stay until the end. If he attempted to do so, it would seem as though he were looking for work, especially when relatives often let words escape that, in a polite way, make his departure expected. In such a case the physician is excusable but not if he is requested not to abandon the patient, for he owes this kindness both to the patient and to the attending friends, even if he is most certain to be of no use at all.

CHAPTER IX
AGAINST THOSE WHO MEASURE THE ABILITY OF A PHYSICIAN BY THE SUCCESS HE HAS, WHICH IS OFTEN DUE TO LUCK MORE THAN TO KNOWLEDGE

It is a strange thing that the science of medicine is deeper and more difficult than any other, yet the dullest of idiots does not fail to pass judgment on the knowledge of physicians. In order to judge someone's ability soundly and fairly, one must at the very least belong to the profession and know something about it. Those who understand nothing of medicine show great temerity when they try to decide who the most knowledgeable physicians are. They base their deliberations upon the success of their techniques; and if someone recovers (even unexpectedly, as we have mentioned above) it is very often attributed to the physician, even though he did nothing of importance. Or the physician is said to do next to nothing when the patient dies or languishes in an illness thought by laymen not to be serious. The more modest laymen, rather than saying that the physician is more or less knowledgeable (if he is considered learned among people of education), will say he is not lucky with his patients and therefore not a good physician, thus still judging on the basis of his success.

It is of course true that in all things there is good and bad luck, and (as is said in Italian) *La buona, e la mala sorte*. And good fortune from the physician's point of view is not to be called or asked for an opinion by patients who end up dying, for no reputation is gained therein, and still less gratitude and friendship. Nevertheless, blaming the physician is not the thing to do, and provided he has done his duty properly, he should not be less esteemed than if the patient had recovered. A captain who has defended a stronghold with his last efforts, having eaten all the horses, mules, dogs, cats, and rats of the besieged city, along with the leather, parchment, and other awful foods (as it was said of those in Sancerre in the year 1573, who ate all objects made of leather and parchment, even the roof slates, out of which they made bread—I do not know how), having lost most of his men, the wall in ruins, with nothing left for support, and forced in the end to hand over the place, deserves no less praise (if not more) than another who saves a well-protected stronghold furnished with all the necessary things that make saving it easy and painless.[1]

This is easy to understand if one has the least bit of common sense and is not hampered by emotions. But most men, who are blinded by them, are unable to realize that it is not the physician's fault when a patient who is dear to them does not recover as they had desired and hoped. Likewise, there is always some rancor and unhappiness toward the captain or governor of a place who must surrender because he did not have enough foresight in the matters of the siege, and in several particulars, right down to the possible use of fire by the attackers.[2]

Yet anyone who has been successful in his undertaking is considered valiant (even if he is the biggest coward in the world). It is truly a great thing to be fortunate in one's affairs, but luck does not have anything to do with knowledge or skill. It is a special gift of God to be called upon to treat those who are destined to recover, those whom

He wishes to keep in health and in whom He allows the virtues of the remedies used to be effective. And it is a curse to be called by those who are to die and whom nothing will help.

Thus, it is improper to determine the ability of physicians by their success, which is due more to chance and to the grace of God than to the skill of man. One must nonetheless not infer or conclude from this that it makes no difference which physician one calls. Nor must one say that if God wants the patient to recover, He will bestow His blessings on the remedies of even the most ignorant physician and will make him fortunate. For although this is possible, it is tempting God, as we have pointed out in the fourth chapter.[3] It is like wanting Him to make bread out of stones, and out of an improper remedy a good one.[4] It is commonly said, "God helps those who help themselves." One must search out the best means one can and leave the rest to God, who has all things in His hands.

CHAPTER X
AGAINST THOSE FOR WHOM EVERYTHING IS SUSPICIOUS AND WHO CALUMNIATE PHYSICIANS FOR NEARLY ALL THAT HAPPENS TO THEIR PATIENTS

Among the greatest sufferings the generous and good-hearted physician has to bear are the reproaches and false accusations of patients and attendants who are so unreasonable as to attribute all the complications that arise in the patient to the remedies prescribed. As to the successes, they doubt that they are at all attributable to the physician.

For when they first see that the patient is very weak, they blame the abstinence or the paucity of nourishment ordered by the physician, or they reproach the bloodletting or the purgation. But it is the illness that causes the weakness, not the remedies, which diminish the ill effects of the disease and maintain the patient in greater strength. Without the use of these remedies he would be still weaker. As proof of this point, does one not see in fact that the patients of those who scorn abstinence, bloodletting, and purgation become still weaker? If those who do not use such remedies maintained greater strength than the others, it could be said that the remedies were the cause of the weakness. But, on the contrary, such people indeed weaken, and in the end more of them die than the others.

So it is with other complications unfairly imputed to the remedies, such as vomiting, diarrhea, nausea, thirst, pain, insomnia, raving, and so forth, which arise because of the illness itself, not from the remedies, as ignorant people believe. If after the patient takes something ordered by the physician, or if as soon as it is applied to the patient, he has a spell of vomiting or of diarrhea, that is the cause of it (they think), especially since he did not have this symptom before taking the remedy. After taking this medicine, this syrup, this restorative, this heartwarming potion, or other, he is so sickened he cannot stand it: thirst (they see) overwhelms him more than before.[1] Now, it is true that these symptoms appear after having taken it, but not because of it.[2] This is as poorly argued as saying, "Since it has snowed my clothes are more worn than before, therefore, the snow is the cause of it"; or "Since I ate some of that capon, I have had a headache, colic, or diarrhea, therefore, the capon caused me such ill effects." Poor fools! All that follows does not necessarily issue from what preceded. This diarrhea, this vomiting, nausea, thirst, insomnia, nightmares, and so forth have causes (other than those known to you) that produce such symptoms as they develop. And whatever the physician may know in the line of treatments likely to arrest the course of an illness or prevent or diminish its complications, that illness will nevertheless do damage up to a certain point (called the climax of the illness). But remedies make this come about much more slowly than if the illness were left to run its course. And if thirst, nausea, and other symptoms increase after properly prescribed remedies, be certain that it is because of the illness going out, in spite of these setbacks and this resistance, and that the illness would be still more violent and the symptoms less bearable if nothing had been done about it, as is seen in the experience of those who scorn such remedies.

For if it is true that some die for lack of treatment (which is a maxim accepted by all), they must have more symptoms, and more violent ones, than those who are left without treatment. One must not, then, suspect or calumniate remedies that might be followed by more severe symptoms and say, "Since that frontal salve he has slept less" or "dreamed more" (for the frontal salve is not the cause, but it is the illness, rather, which could not be overpowered), or "After the heartwarming potion he got the hiccups" or "spasms." True enough, but this tail does not belong to that calf. As is said in the popular proverb: "This is from another barrel."[3]

I am not saying that remedies are not sometimes the cause. No, indeed: some are prescribed most inappropriately. But for the sake of argument I am presupposing that the physician is learned, diligent, and concerned. These qualities should be sensed and his actions interpreted positively, through his prescriptions; and one should attribute to the illness or to the express will of God, rather than to his remedies, any complications that arise or worsen matters. For unexpected things will happen that cannot in any way be foreseen so as to be avoided: for instance, just a light laxative might end up causing bleeding because the patient was on the verge of having a bowel movement. The physician who is unable to guess this, especially in a person neither ill nor well (that is to say, a patient who is not confined to bed because he is only a little ill), orders a light laxative. Yet nature will cause an evacuation on her own, knowing there is some need for it. It thereupon happens that after the laxative's effect, nature runs her course also and causes a bowel movement that goes on inordinately and without stopping, because the expulsive virtue, spurred on by the sharp and biting excrement, cannot retain itself, and the matter, being corrosive, scrapes so much where it exits that blood comes out. The laxative will be accused of everything, when it caused but two or three little stools; all the rest is due to an overflowing and is like a torrent of humors accumulated over a long time. Just as one sometimes pulls a stone from a wall and twelve feet of it come tumbling down, so much is it in a ruinous state. A strong wall needs a cannon or a double cannon; a weak one will be widely breached by a fieldpiece. Thus, in order to judge properly the effect of a remedy, its strength must be known, familiar only to the physician, and it must not be judged by its effects. For if during the working of a medicine, or right afterward, one sees coming about what is not in the nature, strength, or range of that medicine, it must not be attributed to it. If a child pushes a staggering drunkard, who suddenly falls down, it is not the push that was so powerful, but the wine that had dulled him, causing him to stumble along, falling down and getting up again.

Similarly, one could from the same comparison make the point that for a very weak patient a light remedy will have the power to make him stumble and fall to the ground. This is why it is better to use another comparison: I use that of giving a flick on the arm to a pregnant woman who, shortly afterward, aborts. Would it be because of the flick on the arm? No, because it was far from the belly and too light. It must, then, be for other reasons that she was weak and happened to abort.

Thus, many things occur that are in no way connected to one another but are simply chance happenings, not due to the cause commonly claimed.

CHAPTER XI
THAT THERE ARE MORE PHYSICIANS THAN OTHER SORTS OF PEOPLE

It is said that the duke of Ferrara, Alfonso d'Este, once asked his subjects in what profession there were the most people.[1] One said shoemakers, the other seamsters, another carpenters, still another sailors, yet another lawyers, and another plowmen. Gonnella, the famous buffoon, said that there were more physicians than other kinds of people and wagered against his master the duke (who rejected the idea entirely) that he would prove it in twenty-four hours.

The next morning Gonnella came out of his house with a nightcap and a bandage around his head and chin, a hat on top, and his coat gathered up over his shoulders. In this getup he took the road to the palace of His Exellency, going through the Avenue of the Angels. The first person he met asked him what was wrong with him. He replied that he had a violent toothache. "Ha, my friend," said the other. "I have the best remedy in the world for that," he continued and told him it. Gonnella took down his name in his notebook, pretending to write down the remedy. Two steps later he found two or three people together who asked him the same question, and each one gave him a cure for it. He wrote their names, as he had done with the first. And continuing on his way leisurely up the avenue, he did not meet a single person who did not give him some remedy, each cure different from the preceding, and each person telling him that it was tried, tested, and guaranteed. He wrote all their names.

When he arrived at the palace courtyard he was surrounded by people (since he was known by everybody) who, after hearing about his toothache, likewise gave him piles of remedies that each said was the best in the world. He thanked them and wrote their names also. When he entered the duke's chamber, His Excellency cried out to him from the other side of the room: "What's wrong with you, Gonnella?" He answered, most pitifully and wretchedly, "A toothache, the cruelest ever." Thereupon His Excellency said to him: "Come, Gonnella, I'll make you something that will get rid of the pain right away, even if the tooth is rotted. Messer Antonio Musa Brassavolo, my own physician, never made up a better remedy.[2] Do this and that and you'll be cured."

Suddenly Gonnella, throwing down his bandages and all his paraphernalia, cried out: "You also, my lord, are a physician! And here is my list of how many others I've found between my house and yours. There are almost two hundred of them—and I only passed through one avenue! I'll wager I could find more than a thousand in this city if I were to go through all the streets. Find me as many people in other professions."

This is a true story and a very pertinent one, for everybody makes medicine his business. Indeed, there are very few people who do not think they know a lot about it—even more than physicians. I leave aside a few surgeons, barbers, apothecaries, attendants, midwives, charlatans, and other quacks such as merchants who, in order to cut in on a portion of the profession, are master meddlers, thinking they know more than Mister Smart, acting haughtily and meddling in healing several ailments with bold assurance, accompanied by overblown promises. I leave them aside, as I said, even though they amount to a considerable number, for there are so many others that it is a shame. Almost everybody is copying the prescriptions of physicians, taking sick people's pulses, examining urine, giving opinions, and ordering the very opposite of what the physicians say.

If there happen to be a few who are more experienced in the matter, I think the number is so small that it would be quicker to write the names of those who are not so presumptuous than to make a list of so many meddlers (an almost infinite undertaking). And a great many of them are so bold as to suggest in front of the physician—right in his presence—that it is necessary to bleed the patient or not to, and when he is bled that it must be so much blood, that it is not a good idea to purge him because the weather is not right, that the patient must be better nourished, that he needs restoratives, linden tea, consommés, broths, barley porridge, almonds; and is the patient allowed too much liberty, or is he being bothered too much?

In short, the great meddler—the first and foremost judge of all—is the vulgar, injurious, and impious ignoramus who (as Terence used to say) considers nothing right unless it is done his way.[3] And if his opinion is not followed, he attributes the death of the patient or the length of the illness to what was done otherwise. For he imagines and convinces himself that his is the only way to proceed, all others being erroneous, finding fault with anything and everything done any other way. What haughtiness! In the other professions, which are less complex and difficult, where everything is obvious, the artisan is left to do as he sees fit. In medicine, the most complex of all, and where the layman cannot understand much of anything, each wants to govern in the hurly-burly. This is why often we do not see sicknesses come to a good end in patients who are important people: they are attended by too many meddlers. Those over whom less fuss is made recover better.

CHAPTER XII
THAT IT DOES NOT USUALLY PROFIT PATIENTS TO HAVE SEVERAL PHYSICIANS; BUT THAT ONE PHYSICIAN MUST BE MOST ASSIDUOUS IN TREATING THEM

From what we have said in the preceding chapter, this title could be understood in the sense of laymen acting as physicians, but I mean it here in its proper sense, designating those who are true physicians both by knowledge and by profession. It is very reasonable and necessary to have the advice of several in the difficulties and doubtful aspects of an illness. For (as is commonly said): "Four eyes see better than two"; and that presupposes that all of them see clearly.[1] For one person will fix upon one thing, and another upon another, which are then gathered and collated to the benefit of the patient.

But having several physicians as a matter of course, all responsible for the same patient, is not to his advantage. For at every instance they can contradict one another over the smallest detail or over unimportant things, one against the other, more for ostentation's sake than out of necessity. Pliny has pointed this out very clearly in the first chapter of his twenty-ninth book when he writes: "There is no doubt that these physicians, seeking a reputation through some novelty, are suddenly dealing with our lives. Because from such a practice come miserable protestations from every quarter among patients, claiming that there is not one physician of the same opinion, rather, they all seem to be saying something different. Whence the epitaph on the unhappy sepulcher: 'I am lost for having had many physicians.' It refers to the emperor Hadrian who, upon dying, cried out: 'The multitude of physicians is causing me to perish.' "

Now, the reasons for this abuse are diverse, and first among them is the envy or jealousy that one doctor usually harbors for another, especially one who is more ambitious, avaricious, and unscrupulous than others. For it is common for a potter to be envious of another, according to the old proverb,[2] but even more so the physician, since the honor of having diagnosed, treated, and cured the patient would be stripped from him completely. This is why he will not be willing to accept sharing it with another. I am not speaking of the ambitious and avaricious man who besides being an unbearable detractor is also usually quarrelsome, for there are some that are very modest. But they are still highly jealous of the honor they feel people owe them, and they consider themselves most capable of performing simple, common, and ordinary applications without the unpleasantness of being contradicted. Yet they consent and bend to the desire and pleasure of the patient or his relatives. But this is not to the patient's advantage, as I have tried to point out. For even if we were to have three or four physicians seeing to the recovery of a patient, and even if all of them were very modest, amiable, and learned, many of the most common disadvantages I will now discuss could not be avoided. As for the less common disadvantages, I leave to those who have observed a great number of them to judge how the practice of multiple physicians is inconvenient and even harmful for the poor patients.

First of all, if only one or two physicians are attending, they will be more careful, diligent, and concerned, so they can come out of it with their honor. And if there is only one to bear the entire burden on his shoulders, he will be all the more attentive in the matter, since he does not rely on anyone else and all will fall back on him. This is why if he has an honest heart and is a good man he will strive to do better than if he were accompanied, understanding always (as should be the case) that in any difficulty he will seek advice. The feelings a physician has for the patient are not of small importance; they are, rather, so great that they deserve to be put before all else.

The other disadvantage is that it is not easy for several physicians to be able to meet in order to visit the patient at the same time, for each will usually be called to other patients, emergency cases, or smaller matters. This is why we are often forced to miss an appointment to meet at the patient's home. In this event both the regular physician and the others are hesitant to give an opinion or make a decision, fearing that the absent physician will not agree. They are also afraid that a single opinion might cause the patient or attendants to make a mistake because it differs from a later one, asked for secretly. Sometimes this will only be some small disagreement, like over whether or not a cherry can be eaten, which in itself is not worth discussing; but it is necessary that all be in agreement. This troubles physicians, and often the patients suffer for it.

They also suffer (coming to the third point) because of several little things that the attending and regular physician would see to and remedy from moment to moment. (I say "little things," in themselves, yet very often causing great concern.) But afraid the other physicians might not be happy, he will not dare to perform these simple duties. Because of this the patient undergoes a lot of anguish he would otherwise be spared, such as enduring thirst for a long period of time, being kept too warm, taking too much food and medicine, being forbidden some pleasure or recreation not detrimental to his recovery, and so on.

I will limit myself to arguing against these three disadvantages, usually cited by the majority of physicians, in order to show that it is much better to have only one assiduous physician. The greatest luck a patient can have is to find a good physician who will not budge from his side, according to the advice of Jesus the Son of Sirach, whom we quoted in the first chapter. For with just one visit or two a day (it can be said, by and large), the patient will not receive proper care. A physician must be present to observe peculiarities, which can cause him to modify his opinion hourly concerning the food and other remedies. This is why Celsus speaks aptly when he points out what diligence the physician must show in ordering proper nourishment, as to the times and amounts (one of the most important points in the treatment), for, as he writes, the appropriate food is an excellent remedy and good medicine: "It is, however, always and everywhere necessary to be sure that the attending physician note continually the patient's strength and, as long as it is good, that he prescribe abstinence; when he begins to suspect weakness, he must aid him with food. For it is his duty not to overburden the patient with superfluous matter, nor to further his weakness with hunger, etc. From this it can be seen that several patients cannot be cared for by a single physician and that he (if he understands his art well) is an upright man who will not overcharge his patient. But those who are given over to profit, since there is more to gain with a multitude of people, willingly embrace rules that do not require much care, as in this case. For it is very easy to count days, hours, and visits, even for those who do not often see the patient. He must needs be diligent who wishes to see only what is necessary, and when the patient will be too weak if he does not take nourishment."[3]

This is why, for the good of the patient, it is of very great importance to be always attended by a good physician, both for his diet and for his taking of remedies. For, being present, the physician will advance or delay, will increase or diminish the dosage, and will do many things in a different way than if he only sees the patient at long intervals, as is the practice with laymen.

Thus, it would be better to have a physician with a little less wealth and reputation (and consequently less work) who will be more available and more assiduous. For diligence, vigilance, and careful observation by a normal physician can outweigh a greater knowledge that is not applied in practice.

CHAPTER XIII
AGAINST THOSE WHO COMPLAIN ABOUT THE BRIEF VISITS OF SOME PHYSICIANS

As Democritus pointed out to Hippocrates, in a conversation that Hippocrates conveyed in writing to Damagetes, our lives are full of contradictions.[1] Indeed, what we like now, we will dislike an hour from now. The plowman wants to be a soldier and before long will abandon his first occupation. The merchant may play the gentleman but soon returns to his merchandise.

But the contradiction is still more apparent when one seeks opposites in one and the same thing, such as being a soldier and not being bound to fight in wars, being a landowner and not being subject to lawsuits, having a lot of servants and chambermaids and not being stolen from, living dissolutely and never getting sick. Thus it is with many who wish to have the busiest and the most experienced physicians (on which basis laymen judge them to be the most knowledgeable, as is most often the case, but not always). Yet they then complain about these physicians' brief visits and about seeing them so infrequently.

This is a complaint made commonly about Parisian physicians (the most famous), who in such a large city have so many patients as a matter of course that it is impossible for them to stay long with any one of them. For if a physician has to see twenty patients twice a day, is it not a bit much if he stays with each one a quarter of an hour each time? He cannot do any more. For on his longest day, which will be sixteen hours long, I expect him to begin his calls at five in the morning and continue them until ten, then begin again at twelve and go until five in the evening. That makes ten hours spent in house calls. He needs what time is left to rest, from ten to twelve for his repast and nourishment, and likewise from five to seven in the evening, followed by undisturbed sleep. For if he keeps on day and night it is impossible for him to last very long.

I am willing to go as far as six hours in the morning and six after lunch since going from one house to another, up and down stairways, cuts two hours from the time for twenty house calls. Furthermore, one does not run around in town like a courier, and in summer when the days are long such haste and activity are dangerous because of overheating, perspiration, dehydration, and their attendant problems. There remain, then, around ten hours net when the physician can be at his patients' bedsides under the best of circumstances.

And what does that come to for each of the twenty? If I am counting correctly it is a quarter of an hour per patient, mornings as well as afternoons. Now, it is certain that the most famous physician will, on a given day, have to visit more than thirty patients besides doing some consulting, which requires a much longer stay than a simple house call. It thus follows that each of these calls will only be seven or eight minutes long. For it is necessary to please everybody, and of one who divides himself up among several, each can have but a short time. Thus, the physician scarcely does more than come in and go out, inquiring on the run about the state of the patient, taking the pulse, examining the urine, uttering a word or two about what should be done. And away, on to the next one!

One cannot, in all fairness, blame him for his haste and his cursory visit, since he is unable to do otherwise. And those who call him are well aware of it. Furthermore, if the physician says at times he cannot stop because of the many other patients he must attend to, he will be told: "Sir, you have but to pop in and out, the patient will think himself better only at the sight of you; if he only sees you once a day for a moment he is completely satisfied." Another does likewise, then a third, and a fourth. What do you do then? But someone will say: "Still, you must have some regard for people of quality and must spend more time with a great lord, bishop, abbot, count, baron, president, counselor, treasurer, financier general, and other honorable people who have the wherewithal to appreciate the call and compensate considerably better than the others." The answer to this is that one must do one's duty well with respect to all and must acquit oneself faithfully of one's charge, and that, besides, there are more commendable patients, such as one's close family, friends, acquaintances, and those to whom one is highly indebted. These patients should, in truth and according to common sense and decency, have precedence over the others, whatever great rank they might hold, and those from whom one takes no money, by virtue of the indebtedness mentioned above, require more help and care from the physician than those from whom one expects payment.

It is therefore no small thing to have indebted and endeared to oneself some learned and wise physician who will always have more concern for friendship than grandeur. Besides, most of these important people only know the physician by reputation and are even less known to the physician himself. Since there is no familiarity, friendship, or compelling obligation, this physician will not be more suitable than another who, not quite so busy, could be more attentive and helpful.

But human nature is such that we want the one most in vogue, and each wishes to have him just for himself, which is none other than to want the impossible. And then people complain about hasty visits! If you say, "I am not the poorest, and I can pay as much as another," there are a hundred others who will say as much. What can the physician do other than divide his time up so that each will have a little? But he will always reserve a little extra for those to whom he is indebted, as reason and humanity dictate.[2]

For these reasons it would be better if everyone were advised to limit desire to what can be had: that is, a physician, easy to procure among those considered learned, who does not yet have too much work because, neglected in comparison to the others with more experience, he has not yet become famous.[3] If there is some complication in the illness, one can always call in others for consultation. You can be sure that if the ordinary physician who does the consultation is a shrewd man, he will be quick to understand—after only a few words—what must be done. He will then act as is fitting. This is the best advice that can be given a sick person, whatever he suffers from, for good treatment. And if he has the means of retaining the physician for himself so that he is scarcely ever absent, all the better for the patient, as I have argued in the preceding chapter.

CHAPTER XIV
ON THE IMPORTANCE OF THE PATIENT'S CONFIDENCE IN THE PHYSICIAN

Someone might infer from what I conclude in the above chapter that I am rebuking the desire that many people have to be treated by more famous physicians who, because of their reputation, have good-sized clienteles in the large cities. God forbid that I give that impression. I would be doing wrong to those venerable and excellent persons who by their merit have acquired such great renown; and I would be doing patients wrong if I convinced them not to seek after and have recourse to these physicians in healing their ills. For, on the contrary, if one is able to have them at hand when they are needed, they are the best in the world. I only rebuked the common complaint uttered by those who are unfairly displeased with them for not being around whenever they like. As far as famous physicians are concerned, I still say they are often also the best.

It is granted, then, that those who have such a reputation and are in such demand are also the most learned and expert, prosperous in their practices and well liked by their patients. For otherwise their favor would not last, and their reputation, ill founded, would soon go up in smoke. They are most fitting, apt, and proper to treat the worst diseases and the worthiest persons. They also have, because of the respect for their reputation and first rank among physicians, more success in healing their patients. For the good opinion patients have of their physician gives them a certain confidence that helps them to recover better, and more steadily, under his treatment than under another's.

This is why we say commonly in our schools: "He in whom many trust heals more patients." And it is this strong imagination that has great power to make an impression on us, as I have sufficiently demonstrated in the preface to the second book of the *Treatise on Laughter*.[1] It is a power of the soul, which greatly moves the blood and humors in such a way that, if they work with a strong will and confidence, the powers of nature gather to fight off the disease. And yet one sees great changes in the patient just with the arrival of the devoutly awaited physician.[2] For desire and hope being satisfied, the soul awakens and withstands the disease; very often nature will make a courageous sally, impetuously driving out the sickening substance in what is called a crisis.

On the other hand, if the patient does not like the physician and does not see himself treated as he would wish, the physician will not get very far. In this instance, the patient, saddened and discouraged, will become weaker than he otherwise would. For his dashed spirits have no vigor because of the fear and mistrust that have seized his heart.

There is another benefit to be enjoyed by the patient when he has a physician to his liking, desire, and wishes, and from whom he expects great help: it is that he willingly accepts all that he is told with a trust that the treatment will both relieve and heal him. But from another physician he takes everything with disdain and ill will, which will gain him little if anything. For even if a remedy is the best and most appropriate in the world, if one does not respect it, the stomach gets upset and does not profit as much as if it is taken with a cheerful heart. Wine, capon consommé, and partridge meat are very healthy, delicious, and dainty, as food and drink go.[3] But if someone were to partake of them with an upset stomach, with a poor opinion and dislike of the wine steward or of the cook, it would do no good to eat them feeling such repugnance. And what will be the case with things that are in themselves distasteful and naturally despised, like medicine and other remedies? It is often necessary for a patient to endure several vexations that will try his patience much more (and to his own detriment) if he does not have a high opinion of the physician, or at least confidence in him. For the patient will do for a physician he trusts what he will refuse to do for one he does not.

It is thus not in vain that these poor patients request physicians with a grand reputation, who are commonly held in high esteem, for such physicians have more success in their procedures and precepts. But one must not put confidence in those one cannot obtain to such an extent that one is unable to have confidence in other physicians. One must choose a second and a third to go to in case the first and second refuse. When one chooses from among these, one must give them complete trust, hope, and goodwill, without longing for the others, and must trust above all in God, who gives strength to the medicine according to His will and pleasure.[4] It is just as in a marriage, when girls hope to marry into a fine family. If they cannot have their wish, they must be contented with the average and put all their love and affection in the husband chance gives them. God can give them as much (or more) happiness with a humble companion as with the richest in the world. It is thus that one founds a good household; to do otherwise is to court disaster, just as with the patient who has no confidence in his physician and wishes for another.

CHAPTER XV
AGAINST THOSE WHO WANT PHYSICIANS AND YET DO NOT DO WHAT THEY ORDER

Of the patients calling a physician to their aid, there are several types. Some want a lot of medicine and in great variety, never being drugged enough; others, however, want nothing but a good diet and to be well fed. There are still others between the two who refuse everything resolutely and only conform to what they wish. Some accept everything except clysters. I once saw a Venetian gentleman in Narbonne, ambassador to His Lordship, who used to say on the subject of doctors that when he was ill he would believe things they had to say in the negative but not in the affirmative.[1] He was a strong, healthy, and joyful old man who was returning from Spain, where he had completed the term of his legation with King Philip. By negative things he meant what the physicians forbade him, such as drinking wine, eating fruit, standing in drafts, and the like; the affirmative were such things as taking one's medicine, clysters, juleps, and others ordered by physicians.

What a fine position that is, and which several hold to their great detriment! For they do want a physician but try to find one who will only order what they want. Scarcely do they fall short of imitating this Venetian, who at least is willing to abstain from that which is forbidden him. Most of our patients want to do just the opposite. What use is it to have a physician if one is not intent upon executing his advice for the protection of one's life? What help can he be if one does not wish him to use his weapons? It could be likened to a person who has fallen into a pit and who implores your aid, but who reaches out to you with bound hands and will not consider using them to help himself out.

Some say that the presence of the physician consoles, rejoices, and gives them courage, making them feel the illness give way and their strength grow. There are some who say: "I do some things the physician suggests, at least concerning food and diet, but the medicine, I don't want to hear about it." It is just as if the people of a besieged city called some good captain to their aid and defense, and when he came, they did not want to obey him or follow his orders, saying they were content with his presence and that they were strengthened by it. That was enough for them; let him regulate the food and do the administration, but as concerns the fighting and opening fire, they do not want to hear from him. And what is that but to mock the profession (as they say) and to destroy oneself pointlessly? Doing what the physician says is not folly, for Ecclesiasticus teaches us just the opposite, saying that the wise man will not abhor medicine.[2] Is that so hard to accept? Medicine is valid, and God has ordained it to fight illness. Indeed, because health is enjoyable, one treats it as one does enjoyable things, and because illness is loathsome, it is treated as a loathsome thing. It is true that one must be gentle with sick people and not treat them roughly or (as they say) gruffly.[3] Pain is already so distressing that it causes us to refuse many things, to do without many things, or to change them into more pleasant forms. Perhaps the pain thus disguised will last longer, but most prefer such a method to being harassed and pestered with treatment. Some say to their physicians: "Be as patient as I am." And in fact, several prefer recovering over a longer period of time with drugs than in a few days with surgery. Galen advises us to leave it up to the patient. Still, in those cases where there is no choice and where no other treatment can be properly used, when the urgency requires it, it is necessary to assert oneself and point out to the patient the complications that could arise, so that the physician cannot be accused of a dangerous laxity, ignorance, or disloyalty.

It is, then, unwise not to accommodate oneself to all that the physician orders or to neglect the slightest thing. All too often failure to observe what might seem unimportant causes the illness to advance quickly toward death, just as a city will sometimes be lost because of one sentinel or because of one slight breach that did not seem to be of any importance. Does it take more than a spark to inflame an entire haystack, and from there the entire house, and from the house the entire village? From but a small mistake—either an excess or a lack—a great disorder can very often result.

And what will happen to those who scorn the advice of the physician, when we often have all we can do to save those who do everything we say? They remind me of what Celsus wrote: "Intemperate men set in place of the physician's their own eating hours; others, on the contrary, allow (as a gift) their physicians to set the time, reserving unto themselves the quantity of food allowed. They think they are acting liberally when they let the physician decide all the other things but keep free as to the type of food. As if it were a matter of what was being allowed by the physician and not what was beneficial to the patient, to whom it is grievously harmful every time he does not follow orders with respect to the time, quantity, or type of food taken."[4]

And in fact, it would almost be better not to use medicine at all if one is not willing to do everything that its use dictates. For many drugs can be harmful if all the physician's instructions are not heeded. It often happens that those who are so overparticular at first accept everything in the end, but they cannot be kept from dying as they might have been at first, through the grace of God.[5] Just as those who, besieged, were at first very cool about properly defending themselves and using every means: their mattresses, bales of wool, chests, and other pieces of furniture with which they might fortify; their supplies and money to pay the soldiers well; their arms and very selves to fight courageously. In the end, when they see themselves being crushed, they offer purses and jewelry, right down to their very entrails in order to save themselves. But there are no more remedies that will help them: too late did the Phrygians take account, as the proverb goes.[6] For these reasons, let each plan from the beginning to do willingly what the physician suggests and specifies, without any restrictions or distinctions between affirmatives and negatives. Even then, thanks be to God if one escapes at that price!

CHAPTER XVI
AGAINST THE ABSURD IGNORANCE OF THOSE WHO ACCEPT EVERYTHING THE PHYSICIAN SAYS EXCEPT IN THE MATTER OF THE AMOUNT OF FOOD[1]

I never cease to be amazed at the stupidity and foolishness of laymen in believing physicians and submitting entirely to them in the most complex and most serious matters, and yet being stubborn and contradictory in simple and inconsequential things. For if it is a matter of bloodletting, purging, or, what is more, incisions, cauterization, and other serious operations (even removal of a limb), they give their consent, both for themselves and for their dependents, without resisting in the slightest what was suggested by one or several physicians.

But when it comes to food there is much to be contested, not with respect to the quality (which is still the most important and complex aspect, and in which laymen do not oppose or overrule the physician), but the quantity, in which these dolts think themselves experts and masters, in spite of the physician. For to hear them talk, patients are never sufficiently nourished and almost all of them die of starvation. It is true that if they ate continuously they would never die; but eating too often and too many things at once kills most of them. When I say "eat," I mean "to take food," either solid or liquid (it is all the same to me, provided the food enters the stomach). And is it not a terrible shame that physicians are not believed in this matter, which is the easiest to understand and with which all other points concur?

If physicians were furnishing board to these patients, one might expect them to offer light meals in order to save and thus earn more. But since it costs physicians nothing, what does it matter to them whether the patient eats ten capons a day or the equivalent except that such a practice is harmful to him and that they are saddened by it, wishing nothing more than to be honorable in their treatment.

Do you think that the physician, who understands every aspect of the treatment, even the most complex, does not also understand the correct amount of nourishment? Why is he not believed on this score? If he is not thought to understand this point well, he should not be believed in more abstruse and involved things but should be rejected as ignorant and as knowing less than women;[2] and all the more if one thinks he does know the proper quantity but knowingly is weakening the patient, either to retain him longer (and thus get more money out of him) or to have himself more highly esteemed when he finally brings him back from such a low state.[3] For it would be most wicked to jeopardize in this way the patient's life, as I demonstrated in the third chapter; and I would consider quite mad anybody who, having such an opinion of a physician, would call upon him to aid him or his dependents.

But I am sure that there is not a physician in the world who would not be overjoyed at healing his patients within three days, nay, even as soon as he touched or looked upon them, and who could cure every sickness with nothing but food or feasting, doing without any drugs whatsoever. Good God, what a clientele such a doctor would have! He would earn more in one day than others do in an entire year when it became known that he healed sooner and that he had cooks for druggists.

It is therefore a great wrong to attempt to contest or resist the physician who is recognized as learned, prudent, diligent, and well motivated, and who, in addition, wishes for the profit of the patient and wants to be honorable in his practice. It is wrong to contest or question him in the quantity of food as much as in other matters, since he should know how to regulate diet according to the seriousness of the illness and the strength of the patient. In these matters the layman understands nothing and yet, overconfident, thinks he knows better than all the physicians in the world how often, when, and how much food to give the patient. In all other matters the physician is, however, sufficiently credited.

There can be no objection on this point, unless it is that the physician is not always present. For then he cannot take note as well as those present of the weakness of the patient and of his need for food and sustenance. This objection would have some justification if one did not tax in the same fashion the physicians who are present and clinical (that is, who do not leave the sickbed). These doctors are assigned to a certain patient, such as one who travels, and accompany the patient everywhere, staying with him all day long and throughout the night if necessary. Such physicians are not more earnest with this type of patient than with any other, even if pressed and solicited constantly to compound bouillon, barley pottage, broth, restoratives, distillates, etc., in order not to leave the stomach idle for even the slightest moment.

The others, who visit the patient only two or three times a day at the most, cannot properly limit the quantity of food or the times of the meals. This is fittingly done by somebody present, as Celsus rightly points out: "It is always and everywhere necessary to note that the attending physician keep close watch on the strength of the patient and, as long as it is sufficient, combat the illness through abstinence. If he begins to suspect weakness, let him assist with food. . . . Hence it can be seen why several patients cannot be treated by a single physician, and that one who, knowing his art, acts properly in not absenting himself from the patient. But those who seek gain, since it comes from having a large crowd, embrace those precepts that do not require sedulousness, as would be the case here. For counting days and fits is easy, even for those who see the patient infrequently. It is necessary for the physician to be present if he is to see the one thing that matters: when the patient is about to weaken if he does not take nourishment."[4]

But be that as it may, the prudent and learned physician who sees his patients once or twice a day and evaluates carefully the nature of the illness and the strength of the patient will better control the type of food than the wisest or (better) the most overconfident and presumptuous ordinary woman in the world. And if there arrives or befalls the patient some unexpected incident that seems to require more nourishment than the physician had ordered, one has only to notify him, and he will see to it. Or if one has overstepped his orders, thinking to do well in a particular instance, at least let it not be hidden from the physician, so it might be known whether one should continue thus or do otherwise. For if the physician does not know something that was done to the patient, his course of action will not do as much good. He is pulling from one side and the others from the other side, which is often the reason why the physician's hopes, along with the patient's and the family's, are dashed. Women will often feed patients without the physician's knowledge, beyond what he had ordered or thought proper, and not only in amount but type also. Then if his recovery follows, they chirp endlessly and brag shamelessly that the patient would have died if they had listened to the physician. The poor, crazed, harebrained fools do not realize that the patient would have recovered sooner, and that they put him in such mortal danger that, if nature had not been strong enough to fight off their unruliness, the patient would have languished under the burden.

Their troublesomeness is without doubt the reason for the length of several illnesses (I do not say for the death of several, so as not to accuse them of homicide), because it is so very often necessary to repurge the patient.[5] For those who are overnourished accumulate large amounts of excrement, which makes illnesses linger and forces the physician to do frequent purgations. Women think they are strengthening the patient through a lot of food. Hippocrates rebukes them: "The more one nourishes an unclean body, the more one harms it"; and "Food given to a person with fever causes weakness." Still, they think they know more in the matter than all the physicians who ever were. If Aesculapius himself were to come back, he would be believed in everything except in the amount of food and the hours for nourishment, in which matter women have usurped the knowledge, the high jurisdiction, and the last resort.[6]

This is why whoever wishes to be well liked by women and to gather a large clientele must be an advocate or proctor in their court and plead always for excess. Such physicians, very much in demand, are considered the most able and renowned as friends of nature. They are never suspected of causing the death of a patient.[7] For laymen have fewer regrets over the death of those who were amply nourished and given lots of pottage, as if this were the only (or principal) means of keeping the soul in the body. And so when one speaks to them of making consommé, broth, strained meat, jelly, distillates[8] or bouillon, restoratives, and other highly nourishing things (unknown or unused by the ancients, even though they were the fathers of medicine), their ears perk up and they are more than ready to prepare them. Unfortunately, they are not content in having on hand such pleasant and delicious aliments that, in small quantity, nourish and sustain inestimably the weak person who digests poorly. They want, in addition, the patient to eat constantly and in large quantities, so that within twenty-four hours he will have been given the equivalent of three or four capons.

Is this not a most excessive abuse of such food, prescribed for those who have a weak stomach and who spend as much energy digesting this small amount of consommé, broth, or strained meat, etc., as they would in good health digesting heavy foods in large quantity? It is not necessary to give so much at once, nor as frequently, so that the stomach can digest it. Otherwise, everything turns bad for there being too much food or too little time, so the body is deprived of good nourishment and weakens constantly, and illness takes the upper hand. An entire capon, with a slab of veal or a shoulder of lamb, is reduced to a dishful of bouillon, consommé, or distillate. Does not the very weak and feeble patient have enough with just half of this to begin with and the other half six hours later? Is it not as much as if he ate each time half a capon and a proportionate amount of veal or lamb? There is no denying that the dregs or remains of the meat, which become excrement, would end up in the privy. That part is reduced and separated and has no function; the bouillon in this procedure receives only the nourishing extract that is converted into beneficial liquids as nourishment for the entire body. This extract, nevertheless, must be digested by the ailing body, just as the meat of half a capon must be digested by a healthy body. This is why it is important to allow enough time for the body to profit from it, as is required by its weak state. Otherwise, the body labors in vain, and this light food is converted into excrement by indigestion. Hence the illness is maintained, and nothing should be done other than purging and repurging the body.

Let, therefore, women be warned once and for all to believe and to obey physicians, not less in the quantity of food than in the quality, as in all other matters of treatment, since the physician knows all these things thoroughly if he understands the smallest part of his art. There is no physician so nasty and cross that he does not wish to be honorable in his enterprise.

CHAPTER XVII
ON THOSE WHO IN THEIR ILLNESS WANT NO MEDICINE OTHER THAN THAT WHICH ALLEVIATES PAIN

I remember the words of a gentleman from Vivarais whose pleasures meant a great deal to him. He did not pay much attention to illnesses that were painless and thought that medicine did very little if anything for them, as if it were necessary for the illness to run its course. Whatever was done, the sickness would go through its four stages, if it were curable; if it were fatal, nothing could be done. All of these opinions are false, founded upon errors refuted above.

In short, he did not want any physician at all, nor any medicine other than that given to take away pain. But if he had been struck with paralysis, which is a painless illness, I think he would well have wished to remedy it with medicine. As for painful illnesses, it must be understood that pain is not the main consideration (even though an important one), and that pain must be removed at the source if one wishes to give proper treatment.[1]

If one spends time merely alleviating pain and neglects its cause (which is the illness, source, root, and matrix of the pain), there are but two methods: soothing drugs, which alleviate the pain somewhat and make the person able to put up with what remains, and stupefacient drugs, that is, benumbing ones that put the limb to sleep by arresting natural heat.[2] The latter should only be used in dire necessity and with prudence.

But both of these only lighten the pain or make it disappear for a while. One must always come back to treating the essential. Otherwise, one must begin all over again. That our drugs do not have the function of curing diseases (both painless and causing pain) is the greatest falsehood in the world, as I sufficiently demonstrated above when I overturned the argument that physicians serve no useful purpose and do nothing but abuse the populace.[3] If I am reminded by some that several illnesses are healed very well without physicians and without drugs, I will remind them likewise that several patients get rid of their pain without physicians and without any drugs, to the point that such a claim actually dismantles itself.

CHAPTER XVIII
THAT THOSE WHO ARE SUBJECT TO ILLNESS ARE SUBJECT TO MEDICINE; THE OTHERS ARE NOT

Several people reprimand those who follow a diet and take certain medicines in order to keep themselves in good health and prevent the illnesses to which they are susceptible. Those who take such measures are of course very healthy and have a strong constitution. Hence, as far as they are concerned, the proposition is quite true following what is said in Holy Scripture: "The law is not made for the just man."[1] And even more so where it says: "The healthy have no need of a physician."[2] But this statement also confirms the contrary, that is, the ill are in need of a physician, and one who is subject to a certain illness is subject to certain rules in just the same way as we who are subject to sin are subject to the law.

I will always agree with the most eloquent Celsus who says: "The healthy man, while he is feeling well, belongs to himself and does not have to follow any rule or diet nor consult a physician.[3] He must have a varied manner of living, now in the country, now in the city, but more often in the country, sailing, hunting, resting at times but exercising more often. For idleness and laziness make the body blockish; work toughens it. The former ushers in old age, the latter prolongs youth. It is also good to bathe occasionally, sometimes using cold water, at times anointing oneself, at times doing without, fearing no kind of food people normally eat. Sometimes enjoy a feast, sometimes withdraw from it, eating now beyond all bounds, now soberly. Have two meals a day more often than just one, and always eat a lot as long as you are able to digest, etc. [. . .] As for carnal copulation, it must not be overly desired nor overly feared either. When it is infrequent it stirs up the body, when frequent it relaxes it, etc. [. . .] This must be observed by those whose health is sound; and watch out that the remedies for the sick are not taken by the healthy."[4]

Thus, people who are healthy need not pay attention to anything, need not take anything when they are feeling well and their health is sound, as Celsus prescribes. For one would do great harm by making oneself tender and weak, softening and enervating a sound and strong constitution, which gets stronger through all kinds of exercise. But who questions that those who are sickly, unhealthy, and susceptible to certain illnesses, such as epilepsy (commonly called the falling sickness), megrim, rheum, catarrh, shortness of breath, upset stomach, obstruction of the liver or of the spleen, windiness, kidney stones, gout, and similar illnesses (most of which are hereditary, as is leprosy), ought to follow certain rules if they want to be comfortable and live long?

Those who give themselves to study or to public service, since they are subject to a lot of pressure, must follow a regimen or they often become ill. For they force themselves to do many things that are harmful. And Celsus in the matter at hand also supposes that the healthy man belongs entirely to himself. But in our chapter heading when we say "subject to illness" we mean a particular subjection and susceptibility. For all the men in the world are subject to all kinds of illnesses, as they are subject to death. But we say that those who are particularly subject to them are those who have an inclination and disposition to some illness, of which the seed or the rudiment is in them; not that they are in fact ill, but that over nothing they fall ill. And for this reason they must take good care of themselves, like the man we discussed in the second chapter of this book, who, although the unhealthiest person of his time, lived to be a hundred through his extremely careful and fastidious manner of living.

CHAPTER XIX
THAT THOSE WHO KNOW A LITTLE ABOUT MEDICINE ARE WORSE FOR THE SICK THAN THOSE WHO KNOW NOTHING AT ALL

This error ought to have been discussed after the one in the ninth chapter, where I showed that there were more physicians than other sorts of people. But fearful of offending those persons who are very helpful, I was a long time debating whether I should thus publicly disgrace and reprimand them.[1] In the end I became convinced I should proceed, knowing there is more danger than commonly believed in those who think they know something and are confident they know everything.[2]

Out of ignorance, these overconfident people either presume to undertake the gravest of matters, or they resist and hinder physicians from using the main remedies necessary for prompt and sure treatment. Thus these comptrollers hold us checked in fear, so that we dare not do anything while they play haughty.

There are people who know nothing at all about medicine nor anything about arguments or reasoning (such are ignorant women). They do not even know how to read or write but impose on us their observations and their rules. Just because they know how to do a pottage, a sling, restoratives, a barley soup, a proper bed, a patient's hair, and a few simple remedies for the mange, burns, the falling of the uvula, worms, the suffocation of the uterus, etc., they think they know everything and undertake numerous cures according to their own humor and fantasy without the physician's knowing it. But if things go poorly, I do not think you will see them bragging about it. The ample robe of the physician will cover it all.[3]

It would be much better if people attending the patient knew absolutely nothing other than how to follow the physician's orders. This is a course of action most helpful to the patient, for he who is not presumptuous will only undertake what is prescribed, ordered, and commanded. Others, who think they know something, add to the orders, take away from them, change them, or do not follow them at all, like bad apothecaries who execute as they please the physician's prescriptions, thinking they know better the condition of the patient or the nature of the illness, drunk with a high opinion of themselves for having seen several such illnesses, frequented diverse physicians, and observed the success of similar preparations.

Oh dangerous pride! This is the ruin of most physicians. It would be much better, by God, to know nothing at all than to know how to treat as an empiric.[4] Oh what a calamity for the life of the patient and the honor of the physician to have such an arrogant, bold, scheming apothecary. In Italy and Spain also (as I understand it), patients are much better off, for the apothecary does not visit the patient except out of courtesy and friendship, not as an apothecary. And the physicians do not write at the bottom of their prescriptions what the medicine is for, so that the apothecary knows as little about the physician's intent as if he had seen nothing at all. In this way he is unable to abuse the physician's prescriptions or does so far less than our own apothecaries, to whom everything is too plainly divulged.[5]

After the apothecaries (I speak of the bad ones and not the good, prudent, modest, and proper ones who do not meddle in anything other than their own business) the most dangerous people are the patients' guardians or maids, who think they know more than the physician (especially if they have been a long time in service) concerning food, although it is of immeasurable importance in its type, timing, and amount. It is true that as to quality they know as well as the physician; but as to the hours and the amount, they do just as they please. I leave aside the adulterants they use covertly and the omissions they make in our orders. In short, they dispense with everything and act according to their fantasies, and they treat the patient just as casually. Such people are most dangerous. It would be much better to have people who know absolutely nothing and whose only talent is to obey.[6]

CHAPTER XX
CONTAINING FIFTEEN ERRORS; ON THE INGRATITUDE OF PATIENTS TOWARD APOTHECARIES; AND WHY IT IS THAT THEY ARE MOST OFTEN POORLY PAID

The ingratitude of patients toward apothecaries is not the same as toward physicians; it is very different and there can be no just comparison, since physicians, properly speaking, cannot be paid, as I have shown in the fifth chapter, while apothecaries can. For what one gives the apothecary is for his merchandise and for his service, to which monetary value can be assigned. They can be paid a reasonable amount without their expecting further gratitude or debt, except for the thanks always owed a public servant who has acquitted himself loyally and well of his charge in his trade, as one is grateful to a domestic servant who has served faithfully and well, beyond that which was properly paid him in wages. Otherwise, the master is an ingrate, unaware of the truth. For man, being by nature free, is not bound to serve in order to earn money alone, but if he has a good heart he aims to gain, beyond the money, the heart, friendship, and good graces of the one he serves, and he must place these in higher esteem than all else. Reciprocally, the master must always love the one who has served him well and keep him always as a domestic, free to enter his house as long as he lives.

Now, the ingratitude of many people toward apothecaries is usually far out of proportion to the service these poor folks have rendered. For patients get angry at paying such large sums. And then they say: "Their preparations are nothing but herbs and roots that the apothecary has taken from our gardens, meadows, vineyards, fields, or plains. He had little trouble in gathering them; he used a little wood or charcoal in concocting them, and that is the whole of it. He gave me a few clysters, he stayed up two or three nights, he had to get out of bed five or six times. What could all that be worth?

He asks a hundred pounds for the two or three weeks I was ill. Am I really being charged five or six pounds a day for those sloppy soups? And then so much for capons and other food? I gave so much to the physicians and to the surgeon. The guardian or little attendant cost me so much. Good heavens! This illness will end up costing me more than a hundred crowns!"[1]

But it is far worse should the patient happen to die. For then the one who must pay will certainly launch a complaint, saying that the apothecary's drugs were weak, that they were old, and that if the patient had been properly and faithfully cared for he would not have died. And so the apothecary is more often a debtor than a creditor. Some people are so ill-advised as to say: "The physician and the apothecary knew and could see very well that he could not be cured, that he would inevitably die. Why did they make him spend all his money? It is an affront and a swindle! The apothecary ought to lose out, just like one who furnishes money or merchandise to a prodigal child."

Oh, poor people! If you knew for certain that this patient was to die in three or four days, would you cease nourishing him? Do you think it proper to abandon somebody ill, considering the large number of those who recover beyond all hope? It is certain (my friend) that illnesses cost more than they are worth, besides the fact that they ruin the body. But the recovery of health cannot be valued in terms of money, as I have demonstrated in the fifth chapter.

It is true that the apothecary and the guardian can be well paid, just like chicken merchants and other shop owners, servants, or chambermaids. But the preparations and the services of the apothecary should be more handsomely recompensed in view of the pressure and the tedious and inordinate work that the patient necessitates. There are pleasures and services that ought to be purchased at twice the price of others because they are rendered at considerable risk or great inconvenience on the part of those doing them, or with such great need on the part of the one who receives them that their price ought to be increased. Would you not willingly give more for a glass of cold water if you were out in the fields than for a pint of wine in a convenient place? This is because you value much more that which is useful to you when you are in need in order to avoid a great discomfort. Thus, when one is ill, one is not at all concerned with the cost and worth of the drugs; one wants every measure to be taken in such a dire circumstance.

Yet, upon settlement, one meets with ingratitude. And not only is it unlikely that he will be overpaid but also that he will even break even. For it is well known that he uses, along with the herbs and roots of his region, several drugs imported from east and west (far-away countries)—very expensive drugs on which he often loses money, just as other merchants lose on what they buy for resale.

Even if the apothecary were not to lose anything, this still is not enough. Is it not right that he profit from his diligence in preparing and administering remedies? He ought to be better paid than a grocer, because a grocer sells things just as he buys them, without any preparation and without changing their form. The apothecary goes to the trouble of preparing them in a thousand different ways, making them suitable to each patient, just as carpenters and joiners work wood and fashion it into buildings and furniture. The lumber merchant sells it all rough and without form. What is worth more (in your opinion), a huge trunk of a walnut tree or the tables, chairs, and beds that are made from them? These pieces of furniture are worth five or six times more than the material, because of the labor and the skill involved in fashioning them and making them apt for use by man. So, too, the drugs that the grocer sells to the apothecary, prepared according to the physician's prescription, are worth five or six times the cost of the ingredients.[2] And yet people do not want the apothecary to have more than the grocer. One pound of rhubarb will cost nine or ten pounds, more often the higher than the lower price. And I know well that if the apothecary does not get more than fifty pounds for it he will not make enough. And then people want him to give it up for the price he paid for it, as if one would expect furniture for the price of wood.[3] It would be interesting to see a spinet, a lute, a harp, and other musical instruments priced only at the cost of the wood; a key and a lock not more than the brute iron; a clock for the cost of its materials; a painting at the cost of the paint used on it!

And the apothecary who does work far more valuable and more useful will have nothing for his skill and preparation of these drugs but will only be given what he paid for them in their rough form? Well understood! When people are in fact ill they are of another opinion: it seems they cannot pay enough for the medicine and the services of the apothecary. But when they have recovered, they have the price reduced by a fifth or a fourth; some are accustomed to reducing it by a third and others by a half. In this they are most gravely mistaken, thinking they have come out ahead.[4] For if it is the custom of the city or of the household, and (as you would say) a common and accepted practice on the part of apothecaries, they make it all come down to the same thing. For if your custom is—and you are thus in agreement—that a fourth will always be discounted, the apothecary (who maintains that forty-six pounds, five shillings, three pence are justly due him) will raise his bill to the sum of sixty-one pounds, thirteen shillings, eight pence, knowing that you will reduce it by fifteen pounds, eight shillings, five pence, which is one fourth.[5] Thus, there will remain for him the forty-six pounds, five shillings, and three pence he claims are justly due him.

Others have the physicians' bills taxed.[6] What do physicians know of the price of the apothecary's drugs except what he tells them? Do they go to market to buy rhubarb, cassia, manna, guaiacum, quicksilver, musk, amber, and other ingredients? They know nothing of the price of these things but by hearsay from the apothecary, who will tell you just as much as the physicians. Moreover, merchandise often changes in price; hence, a physician who would always keep to the same rate would be underpaying the apothecary or the person who is to pay him. Does one not see how sugar goes up and down in value fairly often? And wax, faggots, saffron, and other merchandise called *Latin*, rather improperly, since they are asked for in French and not in Latin, as we do in the case of drugs for the ill.[7] For all our recipes and prescriptions are in Latin, and yet they are not called Latin merchandise. Since everything changes price at least every four or five years, just as do wheat, wine, oil, linen, wool and silk cloth, lead, copper, gold, silver, etc., how do you expect a physician to charge fairly for an apothecary's merchandise if he always estimates the price of a clyster to be seven and a half shillings, a heart potion twenty shillings, a rhubarb remedy thirty shillings, etc., without dealing unfairly with the merchant or the payer? And what need is there to haggle if he has served you well and faithfully (as should be supposed)? Pay him liberally and he will be more willing to serve you loyally and diligently.

But people do far worse. To begin with, rather than paying, they go from one to another, and most often to a newcomer (on the pretext that his drugs must be stronger), who, in order to have more customers, is happy to receive everybody.[8] And then when it comes to paying, they have a better price because they are starting their practice. But when this newcomer has enough clientele and asks as much as another, people go on to a third. This is the true portrait of poor payers. And could one better spend one's money on whatever merchandise there is, or that is more precious, than on health-restoring medicine, which Herophilus has rightly called the hands of God?[9] And do you reckon as nothing the kindness the apothecary has extended in delivering his goods and in helping you in such necessity? If any loan deserves usury (which should more honestly be called interest or profit),[10] it is that one. Still, one makes long entreaties to get the principal paid. Oh, vile and wretched ingratitude! I am amazed (to come back to my comparison of grocers and apothecaries) that there are more apothecaries than grocers, seeing that the latter earn much more without so much pain and difficulty, besides the fact that their profit is clear and assured.[11]

But what do you think is the source of this wrong, and why are our apothecaries so poorly paid? It is because of their arrogance most often, and because they want to have too many irons in the fire. It seems to them that their reputation depends on having more clients than the others, and on knowing as much as a physician. These are the two follies that cause their downfall, for God's sake, and their great ruin![12] And first, as for wanting to have too many irons in the fire in order to have a grand reputation, that causes them huge debts. For they dare not refuse any one person so as not to reject the others, not even those who they know have left their apothecaries and have failed to pay their bills, as I said.

It is a very good deed not to send away the poor. They should be the first to be cared for and helped through charity, even without our expecting from them any money or profit; we should even give them the fee as a charitable alms. But when it comes to the rich, that is another story. They have the wherewithal to pay for everything in cash, for the most part, and they have better credit than the poor because people want to keep them as clients; and apothecaries can maintain their own reputations by being apothecaries to the best houses or the greatest number. But these rich people in the end often pay only in dismissals, outrages, and reproaches, or even sometimes in threats or beatings. At the very least it is necessary to litigate, in which case the lawyers end up getting more than the merchants. The apothecary's downfall? He wanted to have a good reputation; or, so as not to lose some small amount that was due him, he wanted to go further into the matter, fearing that if he did not continue advancing, another would take his business. He thus gets in over his head, to his great detriment.

The second point of arrogance is when the apothecary wants to be a jack-of-all-trades, outside of his profession and vocation, by contradicting the physician, whose colleague he wishes to be, if not his master and superior.[13] For a few apothecaries seem to think they know more than the most knowledgeable physicians because they have seen numerous ill people and observed the prescriptions of various and sundry physicians who were, in their day, highly regarded. Hence, they think they have acquired through experience a sounder and surer understanding than present-day physicians who prescribe according to contemporary methods. And from this experience, overconfident, they actually dare go about prescribing according to their fantasies, about as apropos as the Magnificat at matins, as the proverb goes.[14]

Stupid louts, and more dangerous than savages! The physicians, who are consummate in their knowledge, are hindered by them. And these cocksure characters go their merry ways. One does them too much honor in calling them empirics, if this word is taken literally.[15] For the Empirics were learned people, constituting a third sect within medicine, no less than the Methodicals and the Dogmatics. Galen recognized these three and then chose to remain with the Dogmatics, or rational, founded on natural reason, upon the perfect knowledge of anatomy, of the elements, composition, powers and faculties, natural, vital, and animal activities, and, in particular, of the true causes and symptoms of diseases, of their natures and diverse circumstances. All of this is unknown to these presumptuous charlatans, arrogant braggarts and dangerous schemers, who have nothing but recipes. It is more than enough that they know their trade and that they exercise it faithfully, following the physicians' prescriptions and orders without adding or subtracting so much as a grain. For to do so is most dangerous.

Let them take note of what their own Saladino tells them in his first article, concerning the interrogations to which the apothecary must submit, where he mentions one who was in the service of the most illustrious king of Aragon, and who was severely punished by him and shamefully exiled in Naples.[16] When the king's physicians prescribed a heart potion in which white coral was an ingredient, this apothecary, not having any, burned some red coral which made it white. When the king became aware of this, he fined the apothecary nine thousand ducats (Nicolas Prevost in telling this story says only one thousand ducats) and henceforth no longer wished him in his service.[17] This is a good example for schemers who not only play at being physicians in ordering all sorts of remedies, but also play the comptroller and change the prescriptions from what was ordered, which is a treachery far more dangerous than counterfeiting money. They therefore would deserve to be cooked alive in boiling oil or have their flesh torn off with pincers, like murderous domestic servants who had cut their masters' throats.

The late Monsieur Torrhilon, first lieutenant in the government and presidial seat of Montpellier, discussing the reformation of the apothecaries, used to tell often about one from Paris who served him in a severe illness when he was in the suburbs of Saint-Germain under the care of one of the most learned physicians of the town. When he recovered, he paid the apothecary's bill handsomely without trying to reduce it in any way, as is done commonly. The apothecary, seeing his honesty, said to him: "Truly, monsieur, I deserve this money. And it was most useful to you that I was your friend. For if I had done exactly as the physician had ordered, you would have died or taken a long time to recover." Then Lord Torrhilon cried out at him saying: "Ha, wicked man! That is why I was ill for so long! If you had done as the physician had ordered, I would have recovered sooner." Thus, this arrogant, presumptuous man, thinking the lord would be still more grateful toward him, lost that which he might have got if he had done his duty. Did he not also deserve to lose the fees he did receive (just as false currency is also confiscated) and, in addition, to be corporally punished? For whoever uses these ruses puts the patient in danger of death or makes him suffer longer, to the ruin of his person and his wealth.

And then they are so blinded by vainglory and false confidence that they dare boast about it after the patient has recovered, that is, when he is no longer in danger of dying from their evil and disloyal procedures.[18] If he had died because of them, the apothecary would not be very hasty to brag about having done otherwise than the physician had ordered. Oh, extreme disloyalty and execrable treachery! This is one of the reasons (in my opinion) why they are so poorly paid. For God does not want them to draw profit from their abuses, but to have every discontent and be paid only rarely, and then through much wrangling. Even this would be too much for those who make gross errors or who wish to play the physician.

"But," one might say, "all of them are not like that and do not thus abuse their trade. These, at least, should be properly paid." This is true, but often the good suffer because of the evil. The number of loyal and faithful ones is so small that no exception is made for them. Everything is put under a general condemnation because of the frequent abuses of the greater number, and up until the time that they are all completely reformed, they will be poorly paid.

Also, several who are most conscientious, dispensing physicians' prescriptions very faithfully and never undertaking anything on their own, fail in another matter which causes them to be poorly paid.[19] It is that they are too easygoing and kind, extending credit longer than they ought to rich and gentle people, more concerned with pleasing them than with losing them. Too much kindness is often harmful and (as is commonly said) turns into subservience. It is reasonable and proper to give pleasure, but it must not come at the price of one's ruin without extenuating circumstances. Rendering good service and wanting to be well paid for it are legitimate things and are closely linked. Hence, he who does not render good service does not deserve to be paid, and he who does not pay well (having the wherewithal, that goes without saying) does not deserve to have good service.

There was an apothecary in Lyons who used to require payment beforehand in cash, which increased the prescriptions people would give to him to be prepared and dispensed. He never did anything any other way, even if his customer was the most trustworthy person in the city. Hence, he was nicknamed the Old Villain. Did he have fewer clients because of it? On the contrary: he was so busy that he could scarcely keep up with seven or eight assistants. For good service was guaranteed by good money. I have no advice to give on the matter, but it seems to me that if all apothecaries did the same with respect to those who are able to pay cash, they would be able to give better service and would thus remain on better terms with clients. For when it becomes necessary to pay a great sum, it is a great vexation.

There is another abuse which causes apothecaries as well as patients a great deal of trouble.[20] It is that they do nothing all day long other than trot from patient to patient, leaving to their assistants the compounding of physicians' prescriptions, as they see fit, to the point that they serve their patients in absentia. Would it not be better to remain in the shop and work, overseeing what his assistants are doing, than to go walking thus from house to house? In Italy and Spain, as I have pointed out in the preceding chapter, apothecaries do not budge from their shops and do not visit patients unless they are family members or close friends, whom they do not visit in their function of apothecary. For they do not bring what the physician ordered, nor do they administer it either. They do not know what usage is to be made of it. Someone from the patient's house comes to get the remedies and the attendants administer them as the physician instructed. The barbers furnish clysters, the attendants administer them. Are such patients any less well cared for? A hundred times better than in France, because the master apothecary, tending to his work and overseeing what his assistants are doing (and not knowing for whom the prescriptions are intended), furnishes more promptly and more faithfully what the physician ordered without losing time wandering about. In Italy and Spain, directions are not specified at the bottom of the prescription, so they do not become empirics and are not so bold as to imitate physicians. But this practice would be bad, one will say, because many patients are much consoled in being visited often by apothecaries, who also happen to know a little something about the art of medicine.

This is the common error of this country, source of a double abuse most harmful to patients.[21] Because, as I have pointed out in the preceding chapter, knowing a little something is dangerous for patients. It is better for them to be cared for and served by people who know nothing other than to obey the physician. As for the visit that is thought to serve as consolation, there is again a double error: one is that the apothecary distracts himself improperly from the principal service of the patient, for whom it would be much better if he did not budge from his shop, as has been said, and if everything were ready by the time the physician had ordered it; the other is that patients deceive themselves greatly in this matter, believing apothecaries to be almost physicians, or the vicars of physicians.

I dare not say (because I am too ashamed) that several patients have more trust in certain apothecaries than in the best physicians in the world. I mean trust, not with respect to the man himself in accomplishing faithfully and well his task (such as not giving poison in the place of a health-giving remedy), but with respect to his command of the art of medicine, because of some claimed experience or observation.[22] What folly! From this it arises most often that the apothecary has the first knowledge of the illness, and is the first to be called in to give his opinion, to make or apply some little remedies. Then if he thinks it necessary, a physician is called in. This is the greatest abuse in the world, which I have refuted elsewhere,[23] demonstrating the danger to which a patient is exposed in consulting an apothecary before a physician and in giving him the first report of his illness, as one would inform some minor ordinary judge of a case that should go to a higher court. It is as if one were pleading before recorders or clerks who know nothing about such cases and who are nothing but aides and administrators of justice.

The other trust that patients have in their apothecaries, that of his not deceiving them in furnishing the proper drugs, is most reasonable. Hence, it is necessary that the apothecary be a most upright man, of good morals and great integrity. For the life of the patient is more in his hands than it is in the physician's. Nonetheless, he should not become proud over this, as do some, saying that patients are more indebted to them than to their physicians, because they are able to do them more good or ill, having the power to give poison instead of restoratives.[24] My friends, never give yourselves over to such bragging, and do not think that patients are thereby more indebted to you than to physicians. For a cook can say as much, or a baker, a cellar master, a pastry cook, a miller, butcher, gardener, cheese maker, fishmonger, chicken merchant, and any other who furnishes or prepares food, whether publicly or privately, right down to the valet or chambermaid who draws water from the well or goes to fetch it at the spring. Do they not all have the means to poison it? And if they do not, do they deserve more thanks or recompense, more honor or reward, than magistrates and other police officials, than stewards or physicians? It is good to abstain from evil, and the upright man abstains from it not out of fear of punishment but because he hates evil and cherishes good. But that merits nothing more than the average reputation of an upright man. For he who abstains from the murder he could commit is not considered as having saved a life. Otherwise, we would be indebted for our lives to everybody around us, inasmuch as it is in their power to cut our throats when we sleep. A pistol is not hard to fire. Every valet and chambermaid can poison us. Just because they do not, does this mean we owe them our lives?

It is most true that the apothecary is able to do it more secretly and in such a way that the death proceeding from his poison will not be attributed to him but rather to the illness. Hence, not doing this evil thing to the patient as he very well could, he will simply have the reputation of an upright man, as he indeed should be. But the patient is not indebted for his life if he was in danger of death, in the same manner that he is indebted to the physician, as I have amply demonstrated in the fifth chapter of this book. The apothecary can take no more credit than the attendant or the cook who made the soups, broths, consommés, and other food. He can take no more credit than those who give the patient water, nor more than those who could have killed him in his sleep or when he was awake but abstained from doing such a horrible thing.

And so, let the apothecary not boast over this, as though deserving more than another from the patient. For there is no comparison whatsoever between him and the physician, who can never be adequately paid, as I have sufficiently proved in the fifth chapter, while the apothecary can be overpaid. Whence he should be happy with an honest and reasonable profit and should not be excessive in exacting payment for what is due him. But most of them actually dare to compare themselves to the physician and say: "Well, the physician will receive in the course of an illness twenty or thirty *écus*, while I will get but twenty or thirty pounds.[25] He furnishes nothing, and I put in my material along with my skills. And in addition, I work much more than the physician."[26]

My friend, it is not thus that one should calculate. The physician has his capital in his mind, acquired through years of study and not at a small price. It is a mental capital, which he distributes little by little to each and every patient according to his particular needs, yet without diminishing it in any way, just as several candles are lit from a single flame, which is not diminished by the sharing. This capital is more worthy and precious than any merchandise in the world; hence, he cannot be paid or sufficiently rewarded with money, as I have concluded in the chapter mentioned above.[27] But *your* capital can be paid and overpaid. As for labor, I wish that the apothecary would work a little more physically, and (if you will) even more mentally, in preparing and dispensing that which is required of him. Does he deserve more recompense for that than the physician? Masons and carpenters working for an architect put in much more labor than he, yet they are paid handsomely enough, happy with ten or twelve shillings a day while the architect earns an *écu*.[28] The captain of a ship works less than the sailors, and yet it is he who earns more. Does the plowman not put in more labor, working for our nourishment, than a painter who portrays a person true to life? Yet the plowman, even though he busies himself with a most necessary thing, will not earn four pounds in eight days, while the painter will have, if he is among the best, twenty-five or thirty *écus* for his work.[29] So, too, the physician, even though he works less, deserves more. For his work is more worthy and such that it cannot be sufficiently rewarded when it is truly propitious. It would be something to see a soldier wanting to be paid a hundred pounds a month by the state, even more than his captain, because he says that he does more work and runs more risks! He is given sentinel duty or the task of being a bodyguard, while the captain is at his ease in a comfortable bed! He must go to skirmishes more frequently than his captain and bear arms and be on foot while the captain has a carriage! Hence, he deserves more wages than the captain. Thus, the enlisted man could well say that he deserves more pay than his master because he has more work and bears arms more often.

And to come back to our subject, the attendant will complain about doing more work than the apothecary while earning less.[30] In short, there is no inferior who does not think he deserves more than those who have the principal charge and responsibility, the tasks most dangerous and less tranquil and secure than the smallest and most abject. The physician has full responsibility (which is very heavy) on his back for the patient. The others who execute his orders have far less work in comparison, if they only realized it. Hence, it is very reasonable that he be more honored, respected, and rewarded.

So, enough! Let the apothecaries be happy with their lot and with the gratitude due them for their faithfulness as good public servants. Let them try to do their duty well toward patients, following the physicians' orders. Let them undertake nothing that is not their trade. Let them not be anxious to have a large clientele but to serve loyally and diligently those who have need of them, content with a fair profit, assiduous in their shops, and careful about loans and advances they make. God will then allow them to be better paid for their investment, skill, and labor.

END OF THE FIRST BOOK

THE SECOND BOOK OF POPULAR ERRORS CONCERNING THE VENEREAL ACT, CONCEPTION, AND GENERATION

CHAPTER I
WHETHER A WOMAN IS ABLE TO CONCEIVE WITHOUT HAVING HAD HER FLOWERS

It is commonly said of women who do not have their natural purgations (and consequently do not have children), "No flowers, no seeds," a similitude taken from plants, which are sterile, bearing neither fruit nor seeds, unless they flower. For the flower is the exordium or foundation (or preparative) for the seed and for the fruit of each plant. The menstrual purgations of women are called "flowers" because they ordinarily precede and prepare for the fruit, which is the child. Hence, it follows that women are unable to produce fruit before they have had their monthly terms.[1] The reason is because sperm received in the womb, and kept there, must immediately nourish itself and grow by consuming the blood of the mother, so it might be strong enough to form a child. Otherwise, conception does not take place.[2]

Now, in order to understand this matter and the marvelous providence of nature, one must realize that woman is of such a constitution and temper that, being colder and more moist than the male, she engenders more blood than she is able to use in nourishing her body. When she attains her twelfth year (which is the upper end of her puberty) and has finished most of her growing, the blood then begins to be superfluous and, not being used up completely in nourishing her body, accumulates little by little around the womb. When there is a considerable quantity of it, it spills forth, out of the body, as something useless.[3] I say useless to the body of the woman or girl, who has for her own needs enough blood that is finer and smoother. For the blood that she thus rejects each month is the least fine and smooth of all her blood, not (as some have thought) infected with a bad and pestilent nature. It is to be reproached only because of its thickness, supposing the woman is otherwise very healthy.

Because she abounds with such blood, nature has ordained the least smooth portion to flow forth each month. And this is nature's great and marvelous providence in making the preparatives for the child. For she has so ordered all things that the female, because of her constitution, accumulates so much blood that from the superfluous portion the conceived sperm can take nourishment and ensure its growth. It is not necessary that it be the most refined and smooth blood. The crudest suffices, especially since the conceived sperm has a great digestive virtue in that it is able to consume such matter. Once the child is formed, his liver is the first to receive it, consuming and making from it very refined blood for the nourishment of the body.

That is how the conception and the gestation of the child has been provided for, his sustenance in the mother's body taken care of by a natural necessity. Thus, it is easy to see that if a woman is short of blood she will not be able to conceive because there is no provision for it around the womb. For as soon as the sperm is lodged in the womb (which is the field of nature), if it does not meet the sanguine humor it requires for its food and sustenance, it flows back out, unable to remain there without being immediately put to work. Hence, even when the entire body of the woman is excessively full of blood, if it is not copious in the area of the womb, or if vessels of the womb are closed or obstructed so that the sperm has no way of being immediately provided with sustenance, all is for nothing.

Before puberty a girl is normally unable to conceive, and afterward too, if she is not capable of having her menses for whatever reason. But is it possible for her to conceive and have a child before this monthly flow has begun? That is the question under discussion in this chapter, to which I respond that it is quite possible, for it can very well happen that at the moment when she should normally begin her flowers, when her blood has accumulated about her womb in order to flow forth in but a few hours, the sperm, received deep in the womb, will remain there, having found its provision available. Through these means the blood will be retained until the child, well nourished and grown, comes into the world. Then that blood, which is superfluous and has not been employed in the sustenance of the child, flows out and is emptied, at least that part which is unnecessary, but what is left over rises suddenly to the breasts in order to be transformed into milk to nourish the newborn child. If the mother begins to nurse the child, she will be able to conceive again immediately, without having had her flowers, that is, without having had her menses. For they are being retained to make milk. But there could be enough blood around the womb to provide food for the sperm, which might find itself there, especially if the suckling child is already getting big and, because it is also eating food, does not suckle as much as it used to.[4] So the menstrual blood does not go to the breasts in the same abundance as before but accumulates around the womb where it has its other retreat.

From that point on, the woman is very likely to become pregnant again, and the child must be weaned.[5] It can also happen that the woman will not even recover from lying-in before she is pregnant again. She will thus have conceived twice without ever having had her flowers, that is, without having spilled forth the superfluous blood from month to month, and she will be able to continue thus all her life, being constantly either pregnant, nursing, or lying-in. It is said that a woman from Montauban had seven children without ever having had her flowers, a most strange thing. But I heard that a woman from around Toulouse, of a hale and hearty constitution, has had eighteen (as many males as females) without ever having had any discharge other than that coming from childbirth. I learned this from Madame la Marechale de Montluc, who says she has a neighbor with the same experience.[6]

And yet, it is necessary to maintain this distinction in answering the question at hand: that a woman is able to conceive without having had her flowers flow forth externally and is not able to conceive without having her flowers, or menstrual blood, ready to flow forth, gathered all about the womb. For it does not flow out in women who are healthy (as we always suppose those to be, of whom we speak generally), unless there is a lack of use for it when there is enough either to nourish the sperm held in the womb or to make milk. It is true that a nursing mother can have her flowers even though she has a lot of milk, because she will have blood in excess, more than can be used in milk, beyond her own nourishment. Thus, it is not necessarily the case that every woman who has her menses will regularly and healthily conceive, for other conditions are required for the conception and generation of children, which, not being the subject of our discussion, I shall pass over in silence.[7]

I have done enough to show how it is understood that a woman is able to have children without having had her flowers. I might add here what is written about women from Brazil (a country also called America, discovered and first known by our peers), who never have menstrual flow, like female animals, whose nature and constitution these Brazilians resemble considerably. But since these poor women, living like savages, completely naked and in great hardship, follow the nature of animals, they should not be considered in the present discussion, especially since it is most likely and believable that, if they were to change their climate and living habits, were transported to cooler countries, clothed, nourished, and housed, changing their makeup, they would also behave like our women and would accumulate excess blood, which, unless it were used to nourish a child, would flow forth once each month.[8]

CHAPTER II
WHETHER IT IS POSSIBLE FOR A GIRL TO CONCEIVE AT NINE OR TEN YEARS OF AGE

The most illustrious prince of Salerno, Ferrante da Sanseverino,[1] recently deceased, told me long ago in the city of Alais, where he was married, that in the countryside near Salerno a nine-year-old girl had a child, and that the child lived. I heard of another in Paris who had a child at ten years of age. It is likewise affirmed (and this is duly witnessed) that in Lectoure, a city in Gascony, a girl had a child at nine years of age. She is still alive, named Jane du Peirie, the wife of Vidau Beghe, who during his lifetime was a receiver of fines for the king of Navarre, in the above place.[2] She aborted a son at the age of nine, then at age eleven she had a girl who lived and also had children. At fourteen she had a son named Laurent who is still alive; at sixteen, another who is still alive, named Pierre. Five years later (which was her twenty-first year) she had a girl, the widow today of an apothecary. After that she ceased bearing children, even though her husband was still alive.

But how can this be if it is true that a woman cannot conceive before having her flowers, either internally or externally, and if she is not able to have children before puberty (when her body begins to have less need of blood, which women produce in great quantity), as we have pointed out in the preceding chapter? Puberty occurs at twelve years of age in females and at fourteen in males,[3] when both begin to have hair around their shameful parts, in the area called *pubes* in Latin and in French *penil*. Hence, the obvious exsiccation of the body and the visible changing of its first constitution is sufficiently in evidence.

What we say happens to females at twelve years of age is the average and the usual, but it is not impossible for it to happen sooner, since there are things in nature that are most rare. It is possible that one girl will at ten years of age be more grown, corpulent, and nourished than another at fifteen or twenty, and even that she will stop growing sooner and be in her puberty, having matured as much at nine or ten as other girls normally at fourteen or twenty. That is not impossible. If one is able to have at a young age apt enough members of copulation (as is possible, given the corpulence of the body) and to have blood in abundance so that the received sperm can be maintained, what obstacle can there be to the girl's conceiving before ten? Age has nothing to do with it; the number is but a counting, and years are nothing more than the ends and the limits of the change in constitution. Hence, if the constitution is such in one person at ten as it is at fifteen in others (as can certainly be the case), and if she has the required weight, it must not be doubted that the rest can come about. Thus it is, as we see, with the mind: there are persons who are as wise, circumspect, sharp, clever, wary, and sound in speech and opinion at fifteen as others are at twenty-five and, consequently, as capable of administration and management of their goods or some other charge.

Now, we say in medicine and moral philosophy that the behavior of the mind follows the temper of the body. Hence, the condition of one can be understood from the other. This is why what is seen as miraculous in the mind can also be seen as astounding in the body, such as conceiving and having children at nine or ten, just as a mind might bring forth beautiful works, orations, poems, and other great compositions at so young an age that it is almost unbelievable. This is how it was with the Spaniard Miguel Verin, who died at eighteen after having composed a collection of moral poetry of great wisdom and brilliance.[4] Thus, what is said of these girls is very plausible for the reasons I have discussed, and is consequently believable, especially when duly witnessed.

Furthermore, it is likely that several girls would similarly conceive before puberty if the experiment were made. But popular belief is to the contrary, and indeed it is most proper to abstain from such a practice for other reasons; it is wisely advised that they not marry the moment they are able to. First of all, young girls do not have the discretion, sense, and judgment to take care of their husbands or entertain them before they are older. Secondly, it could stop them from developing as much as they might; hence, it would result in the human race becoming shorter, for both men and women would remain shorter and engender similar offspring. Moreover, children born of very young mothers and fathers are less healthy, just like those born of very old people. Likewise, very young mothers are in danger of dying in childbirth. The Philosopher adds to these reasons that girls who marry young are more lascivious.[5] This is why he wisely advises us not to marry them off before eighteen, nor men before thirty-six. There will thus be more decent women and good housewives to have more beautiful, taller, and healthier children, as they indeed are when the father and mother, having been well nourished, have stopped growing.

Since writing this I have been to Lectoure to see the woman who had had a child at nine years of age, and I spoke with her about it. She was married when she was only seven or eight to Vidau Beghe (who was more than twenty-five), abandoned by her parents to the entire disposal of her husband (whose case is less miraculous given his age). She is a small woman of medium weight, at the age today (which is the fifth of April, 1577) of forty-four. She told me that since her first child (aborted when she was but nine years old), she had always had her flowers regularly. After twenty-one she no longer became pregnant, although she lived with her husband another nineteen years.

Whoever does not want to believe this narration, duly witnessed by honest people still alive and in considerable number, will scarcely believe that a ten-year-old child has gotten a woman pregnant. For one grants more easily that the female is much sooner ready to conceive than is the male to engender. And yet Saint Jerome (one of the four principal Doctors of the Church) attests in writing to a priest named Vital that a boy of ten got his nurse pregnant and that she continued sleeping with him.[6] On this subject it is commonly said that a small rooster has seed, that is to say, that the child had sperm even though he was small.[7]

CHAPTER III
WHETHER OR NOT THE RED MARKS CHILDREN HAVE FROM BIRTH ARE THERE FROM CONCEPTION; AND IF IT IS POSSIBLE FOR A WOMAN TO CONCEIVE WHILE SHE IS HAVING HER FLOWERS

There are children, both boys and girls, who are born with red marks on their faces, necks, shoulders, and other areas of their bodies. It is said that they result from their mother's having conceived and engendered them while having her flowers, just like those who have knobby and crumbling fingernails.

But I hold that it is impossible that, during her menstrual flow, a woman can conceive. This I have amply demonstrated in the First Paradox of my *Second Decade,*[1] giving as reasons, among others, that the sperm cannot attach itself to the inside of the womb and be retained there while the blood is flowing. For, on the contrary, this blood would carry the sperm away with it like a torrent flooding from every direction. Furthermore, for conception and retention of the sperm, which requires blood immediately for its sustenance, this blood must not be driven out of the womb by the expulsive faculty but must be attracted to the sperm itself, little by little, like the dew, just as the other parts of our bodies do for their nourishment. For if the blood is expelled from the womb violently and in abundance, the area will be swollen with it and will have an inflammation called *phlegmon*, and the sperm will not be nourished, but overwhelmed. Hence, it is not possible that women conceive during their flowers unless toward the end, as Aristotle says, when, being neither copious nor precipitated, the flowers can be stopped and suppressed by the sperm attaching itself to the wall of the womb, like glue, for which reason this blood begins to run thin, attracted little by little by the sperm.[2] This last blood is less coarse and unrefined, for it is always the most useless that flows first. Hence, the last to flow, more akin to normal blood, resembles what remains. This is why pregnancy is more healthy if the woman conceives toward the end of her menses than at the beginning.

But since the sperm is able to suppress the menses as they come to an end, can they cause these red marks? No, in my opinion, because the blood does not move toward the sperm unless it is attracted there. And it is attracted very gently, insofar as the sperm is able to transmute it into itself for its sustenance and growth. The child already formed does likewise. It must not be thought that the blood pounces on one or the other, or that it melts and mixes with the sperm, causing the latter to be spotted by it. That is entirely erroneous. If the blood were to flow about in this manner within the womb, the sperm would be of no use to conception. Thus, these red marks must not be associated with the menstrual blood flowing at the time of conception.

Where then do they come from? They can result from some blow, pressure, or jolting that the mother had on some occasion without paying much attention to it or even thinking much about it. Yet such bruisings do not usually last very long, but either heal or suppurate. Madame de Montluc showed me on her youngest daughter the spot where she had some of these marks for more than a year after she was born.[3] It was on the left shoulder, about the size of a *douzain*.[4] Toward the end, the area suppurated, and the wound took a long time to heal because of irritated tissue, which had to be eaten away or removed with corrosives. Is it not, then, a case of the disordered body being blemished by a red morphew,[5] as can happen long after birth? For our bodies are subject to all sorts of morphews and marks in diverse places, either because of diet or the disturbed constitution of the area in which these marks show up. Why would it not happen thus to the child inside its mother, when it is more fragile and sensitive? Is it not subject to morphews and to other illnesses just as one that is born? The child can for similar reasons be stricken with disease and spotting of the skin.

CHAPTER IV
WHY IT IS THAT A WOMAN CONCEIVING TOWARD THE END OF HER FLOWERS OR SOON THEREAFTER USUALLY BECOMES PREGNANT WITH A SON, AND A WOMAN CONCEIVING JUST BEFORE, WITH A DAUGHTER

The proportions are not universal, nor is it always the case, but it is true most of the time, as the experience of many bears out.[1] It is our task to give the reason for this and, if there is a basis for it, to dwell on this point, since it could be of interest to men who wish to have male offspring, whether for assistance or for the inheritance of property, for the honor and recognition or for the appointments granted to male lines. Even if it is only a matter of the excellence of the male sex, there is reason enough to want male offspring. For one is always more attracted to what is more perfect, either in and of itself, or according to our judgment, opinions, or feelings.

Now, without doubt, the male is more worthy, excellent, and perfect than the female: witness the authority and preeminence God has given him, constituting him as lord and master over his wife. Thus, the female is, as it were, a defect, quite unable to improve. For nature aims always to make her work perfect and entire; but if the matter is not apt for the endeavor, she approaches as closely as she can the perfect. Thus, if the matter is not apt and fitting for the forming of a son, she makes a daughter out of it, which is (as Aristotle says) a mutilated and imperfect male.[2]

One has such wishes because of this natural instinct to have sons rather than daughters, although both are good. Accordingly, it will serve the public to note the following small observation, along with the explanation for it: it is necessary first to suppose that the female, colder and more humid by nature than the male, likes food of the same sort, for each is maintained by that which corresponds to his makeup. Thus, the sperm, being retained in the womb and indifferent to either sex (for sperm is neither masculine nor feminine, but apt to become either sex), will be converted into a male or female body according to the conditions of the womb and of the menstrual blood.[3] Likewise, the wheat and barley seed convert into ray or darnel and others into way bennet, and other seeds degenerate because of rainy weather and superfluous moisture in the ground.[4] So, it is certain that man's sperm, even though capable in itself of making a male, often degenerates in the female because of the cold and the moisture of the womb (which is called nature's field) and because of the overwhelming abundance of thick and crude menstrual blood. This is, incidentally, just the moment at which the woman is to have her flowers, for the womb is then very moist with the humor that lurks all around it like standing water. And, on the contrary, after it has flowed out, the womb becomes dry and warmer, having blood similar to that in the rest of the body. Thus, at this moment the woman is more likely to conceive a son, just as she is more likely to conceive a daughter when her flowers first begin.

We can no longer doubt what I say about sperm: it is indifferent as to the two sexes, and nature always tries to make a male out of it, as is the case when the husband or male furnishes more and better sperm with formative virtues. The sperm of the woman is in doubt, if she has any part in it at all.[5] Thus, the generation of males only would be the normal outcome, just as good wheat begets good wheat if the field is properly conditioned. For it is the field and the overly moist weather that cause the good seeds to degenerate into the bad, or into less good ones. Farmers know very well that seed loses its strength little by little and, in the end, is debased if they plant continually the same field. Because of this they advise changing the seed from time to time, taking some from elsewhere. Just as a woman who had only daughters with her first husband has many sons with her second; or the man who had only daughters from his first wife (because she was altering the constitution of his sperm, making it more cold and moist) has from his second wife many sons. For the field is ready for it and harmonizes completely with the qualities of the husband's sperm.

But it must be understood that very often the nature of the mother's womb and blood is the reason that the sperm of the phlegmatic father (more likely to produce daughters than sons) will convert into a more tempered constitution and become matter for a son.[6] For just as the soil can weaken and corrupt seed, it can also correct imperfections. Thus, one often sees more beautiful fruit from trees when they have been transplanted or sown in a place other than where they first grew. For the new field shares its virtues with them. Likewise, the pure and clean womb, dried of its superfluous humor and warmed up (as it is after the menstrual flow) is more likely to produce a son if the sperm complies with its own nature or is not extremely contrary to it.

CHAPTER V
AGAINST THOSE WHO ADVISE KNOWING THE WOMAN DURING HER FLOWERS SO AS NOT TO FAIL IMPREGNATING HER

This advice is not only unbeseeming and against good manners but is also contrary to the commandment of God, who most expressly forbids it in Leviticus.[1] Women did not even dare go to the temple during their menses, for they were considered unclean. Men who transgressed by knowing them in this state were polluted and befouled. Carnal knowledge with a woman in her flowers was forbidden for good reason: not out of fear that the child conceived during the menses might be leprous or subject to leprosy, as many believe,[2] but because at this time the woman is unlikely to conceive, which is the principal end of copulation, and because it is a foul, indecent, and beastly thing to have to do to a woman when she is thus purging herself.

That it is not out of fear that the child will be a leper, we have sufficiently pointed out in the two preceding chapters. A woman cannot conceive during her flowers. This opinion is thus refuted, as well as advice, which is not only against the law of God and of decency, but also against the law of nature and her designs. One thinks one is impregnating all the better, yet it is impossible (unless it is toward the end of the flowers, as we said in the third chapter, for at that moment it is feasible, but more decently and surely so when she is completely dried up). As we shall demonstrate in the next chapter, a woman who has been purged and cleaned is more likely to conceive. This is why some women become pregnant just after having been purged with remedies because of some present or imminent illness, without pregnancy being the physician's intention, or their own.

But, I ask you, what is the source of this notion of knowing a woman during her flowers in order to impregnate her? It is, in my opinion, only from those who have neither the strength nor the means and ability of thrusting into the mouth of the womb, in which place the encounter between the two sperms occurs for conception and generation.[3] Why? Because this mouth is too narrow for their (possibly) weak member, or because the mouth of the womb is so frothy and slimy that the member, arriving only just that far, slides from side to side without managing to enter.

Now, during the monthly flowers, this passage is more loose, soft, and open, and thus more accessible. But, as I have sufficiently shown, a woman in this state is not apt to conceive. And this is why one abstains from such nastiness. The monk mentioned in the twenty-third story (from the third story of the third day) of the queen of Navarre's *Heptameron* was certainly of this opinion, theologically speaking, but speaking as a doctor to the gentleman of Périgord, his advice not to sleep with his wife until two hours after midnight was for no noble reason.[4] For he himself had no fear of troubling the digestion of the good lady!

CHAPTER VI
AGAINST THOSE WHO NEVER STOP EMBRACING THEIR WIVES IN ORDER TO HAVE CHILDREN; AND AGAINST THOSE WHO DO IT RARELY TO HAVE FEWER

Ignorant people are mistaken in two opposite ways, courses of action which work completely against their intentions. This happens when some, very desirous of having children, never stop embracing their wives as often as they can, while others go easy on them, fearing they will have too large a family. The former think that if they miss on one occasion, the other ones will make up for it; but things happen completely otherwise. For what could be done in one stroke can be undone with a second one. Moreover, when one goes back in so often (even without nature's invitation), the sperm does not have time to be elaborated and perfected; it is not, therefore, fecund and prolific, but as useless as water.

All sperm is not suitable for making children: it must have two necessary qualities. First, there must be enough of it; second, it has to be well concocted and refined, thick, sticky, and full of wriggling spirits. Both of these qualities are lacking in people who go at it so often. For they could be the best-fed people in the world (since it is an occupation that enjoys good living, for Venus is cold without bread and wine, as the proverb goes)[1] and the best rested: it is still impossible for there to be a supply of sperm and for it to be refined.

Others, who sleep with their wives less often, fare better. For the man in the meantime (if he is continent and does not make love elsewhere, that goes without saying) stores up a mass of sperm that slowly becomes perfected in quality so that, the first time, if the woman is properly disposed, he impregnates her, quite contrary to his intention. This is why one sees several women who no sooner recover from lying-in than they are pregnant again, especially since the husband had been storing up matter for three weeks or a month, and the wife has a well-purged womb. She, too, having been better fed than usual (especially if she gives birth to a son, which naturally causes more joy than a daughter), has for her part accumulated a lot of sperm,[2] which arouses her and makes her more lustful for the male than she has been for some time. For during pregnancy, when the womb is full, she takes less pleasure in copulation; but after her lying-in the womb begins crying out in hunger and has a bigger appetite than before. This is why the woman, driven by this lustiness, quickly forgets the vows and protestations she uttered, racked with pain, during childbirth when she had to vomit back the pleasure received previously. At this moment she would much rather never have a child again, desiring to be henceforth sterile and (if it were possible without further pain) no longer to have the organs of copulation.

But after a few days go by, and the soreness in the breasts and the stabbing and gripping pains are over, all is forgotten and the womb begins to wriggle, feeling like some love games, or having even more desperate hunger than ever for the pleasures tasted in the past. All the more so if the woman has been well cared for and given a bath, and other niceties to tauten the belly, tighten up the openings, and fix everything up so that it seems nothing had been touched. Then the woman is most ready to conceive.

One sees something similar upon the husband's return after a trip, when the wife suddenly becomes pregnant because the husband has the wherewithal (if he has been a good husband and not broken his wedding vows) and the wife, having waited a long time, is starved for him. Upon seeing each other again after a long absence, it seems they make love as they did on their wedding night.

From these observations and conclusions drawn from them, it is easy to understand that he who does it often is more certain of impregnating his wife, provided (as I have said) that he does not cheat on her and neglect her for his whores. For this would indeed be a way not to have much of a family if one were to sow in one's field nothing but withered and feeble seed, while the best was used in the cultivation of wild love, in which bad husbands expend the cream of their sound constitution and all their lustiness, keeping for their wives nothing but the coarse bread and the bottom of the barrel. These are mean, infamous, and wicked adulterous people, whom God does not allow to multiply in a beautiful lineage and with legitimate children, true inheritors of their wealth and honors; they fill their houses with weakling bastards who stand before their eyes as reminders of their sin, over whom (if they have any fear of God) they should feel great shame and compunction, with continual repentance, and should groan from the depths of their hearts, as David did.[3] With legitimate children, on the contrary, one gives glory to God, and one rejoices in them openly, sharing among them wealth and honors with great satisfaction.

CHAPTER VII
THAT ONE MUST NOT KNOW THE WOMAN BEFORE GOING TO SLEEP; AND THAT BECAUSE WORKERS DO NOT, THEY ARE LESS GOUTY AND HAVE MORE CHILDREN

I have two things to point out: why it is that workers (such as plowmen and artisans) usually have more children than people who sit or people of means, and why it is that they are less gouty. I am passing over the other causes of gout for the moment; here, where I am discussing generation, it is enough to note that gout very often comes from the importunate and untimely venereal act. That is, when one gives oneself to it before the stomach has finished digesting, after having drunk, as do naturally those who are too much subject and given over to carnal, licentious, and lecherous voluptuousness. For these people, any time is a good one, that is, they do not observe any special hours, but rather, full of idleness (used for a "good time"), overnourished bodily, underfed mentally, they go looking for such business, and force themselves, nay, push and lay hard unto nature with this folly, for which, in the end, they pay dearly. They foreshorten their lives considerably, just as do lascivious and wanton lechers,[1] who do not live very long but fall quickly subject, prey, and victim to gout, colic, nephritis, apoplexy, paralysis, convulsions, and other diseases of indigestion (the cause of phlegm, father of all these diseases). Because the lecher loses large amounts of spirits[2] and natural heat by using up a lot of blood (a substance very close to sperm, and one from which it is produced), the parts responsible for the body's nourishment are made cold and weak and, consequently, cannot allow for proper digestion.

And now for frequency, or excessive duration of the venereal act, indulged in by people who live a life of ease and spend more time on their pleasures than the poor workers (who must think more about what they are going to live on that day than about making love, even if work toughens them and makes them stronger, rendering them less fragile and less subject to disease). The other consideration is one of time, and on this matter we say that the importunate and untimely venereal act is a cause of indigestion and a bad stomach, as when one goes at it right after a meal and as soon as one is in bed, as is naturally the case with people who are idle and who sit.

The poor workers, on the contrary, who are very tired after the day's work, no sooner are in bed than they are asleep; and if they have anything to ask of their wives, it is after having rested, slept, and digested supper. At this time they have more enjoyment, do it better, at their ease, lustily, and get the good one should out of this natural act: namely, they get up more nimble and lively because the natural heat is increased by it, not dissipated or weakened. And they are more certain of impregnating their wives, if there be cause.

This brings us to the other point: the large number of children one notices among workers (more than among the rich and well off). The reason for this can be found in what was shown in the preceding chapters five and six: the longer sperm remains in its vesicles and is not spilled or spread about prodigally, the more it is fecund and prolific. This is what one observes most often and for the most part among the poor, chaste, and continent workers, both with respect to the work, which occupies them, and the poverty, which makes them happy with their ordinary lot. Because they build up a better stock of sperm and use it in a better way, they rarely miss the mark (if the wife is properly disposed). This is how they fill their houses with children, because of whom they are still poorer, but not in that grace and blessing that the royal psalmist David promises those who fear God, who provides everything in His bounty and generosity.[3] This is also how they are less gouty from any venereal cause and, at the same time, father healthier and more robust children.

Now, on the necessity of not knowing the woman before sleeping, and following the example of these good people beyond the very successful experience I have just discussed and for the reasons given, I wish to demonstrate and prove it still further. Wakefulness, an activity of the animal virtues or faculties, causes considerable dissipation of spirits, even in the most idle person in the world, just as the exercise of the external senses (and especially of sight) consumes a lot of spirits. Likewise with speech, and with all movement, calculating, speaking, thinking, along with the emotions, whether joy or laughing, hope or fear, and similar actions or feelings, all of which cause a notable dissipation of spirits and of fine blood, as long as one is awake. Because of this, one is finally forced to sleep, during which the animal functions stop and rest so that during these cessations one might reconstitute one's spirits and amass them so as to provide for another period of wakefulness. Otherwise, lessened and impaired, the body dissolves and decays, because all its food (or most of it) is used up in replenishing spirits to animate wakefulness.

If, then, wakefulness causes dissipation of spirits, which in turn requires and calls for the necessity of sleep (a saving and a withdrawing from this considerable expenditure), and if on another account the venereal act also causes a notable consumption or use of spirits, it is certain that such an act is inappropriate, or (as Celsus says) worse during the day and more sure at night.[4] But this is on the condition that, as the same author prescribes, right afterward one does not attempt to stay awake and to work besides. For after this act it is necessary to rest and even to sleep a little, if possible, so as not to suffer loss upon loss of spirits. Thus, the most appropriate time is after the first sleep, when one has satisfied nature, when one has replenished a good amount of the dissipated and spent spirits of the preceding wakefulness, and when the body has enjoyed the profits of the food taken during the day. This is when one must turn to one's wife, if one is invited by the stirrings of the flesh, and must immediately thereafter get back to sleep again, if possible, or if not, at least to remain in bed and relax while talking together joyfully.

CHAPTER VIII
HOW IT IS TO BE UNDERSTOOD THAT ONE HOUR SOONER OR LATER MAKES FOR THE ENGENDERING OF A SON OR A DAUGHTER

This statement depends once again on the preceding ones, and especially on the one we discussed in the fourth chapter, where we said that sperm is indifferent as to the two sexes. This must be kept in mind; it is various differences in its constitution that makes it more inclined to become one or the other. When it is hot and dry, for example, it naturally turns into a masculine body (if it meets the field so disposed); or because of the alteration that the sperm undergoes in the womb, it will become a female (as if degenerating from the more perfect). If, then, the body of a male requires a more elaborated, hotter, and drier sperm than that of a female, and if such perfection and constitution are acquired through a long stay and a continued elaboration (for the longer the sperm remains in the vesicles, the more it will be said to be refined, thick, sticky, and full of spirits), it follows that those who go at it less often make more males, and that by knowing their wives as soon as they are in bed they are making daughters instead of sons. For such sperm is not at that moment as well provided with everything required for its perfection as it will be the next morning, after having rested longer. Thus, the morning is more appropriate for producing sons, who will, moreover, be lustier and stronger, as we said of those born of poor people.

But (you will say),[1] there could be some sperm in the spermatic parts that has been there longer than that very day. What is more, from what is eaten, sperm cannot be made for at least a whole day, for it takes quite some time to convert food into chyme, then into blood, then into sperm. Thus, there is the need to wait, simply for the stomach to digest, as the monk mentioned at the end of the fifth chapter would have it. This is especially true because, with the food still in the stomach, all the parts of the body feel it a little and are, as it were, refreshed by its vapor.[2] They thus feel strengthened by the food even before it is made into blood for their nourishment. Now, this vapor makes sperm that had already been well elaborated somewhat raw again upon its first encounter with it. This is why it is better to postpone carnal knowledge with one's wife for some time after a meal in order to make good work of it and engender strong sons, as I said those of poor people are.

One must not, however, object by saying that poor people have daughters just like the rich.[3] For the poor do not always observe the above rule of sleeping and declining before joining but are in this matter very disordered, especially on feast days, when most of them go to taverns and spend in one stroke more money than they earn in three days and come home very drunk. If the wife takes too much note of it or reproaches her husband for his merrymaking, she gets a beating, and as soon as they are in bed the good fellow wants to make peace. Or if the wife says not a word, the husband, wishing to share his merrymaking, squeezes her more lovingly than usual. Here, by God, is where their daughters are most often produced. Even if they waited until the next morning, because they got drunk the day before, they would scarcely do much better, unless it be a girl that is stronger, as are those seen to be more manly, lacking only a beard, although not by much.

From these arguments one can understand why we are inclined to say that one hour sooner or later makes for the engendering of a male or a female. We understand by "hour" a certain period of time, not precisely the one-twenty-fourth part of a natural day, even though in this strict sense the statement could still be true. For on occasion it is a matter of very little time for the sperm to reach refinement and perfection, as we see with fruit that is picked a little too soon or a little too late, and with food we cook, and especially with distillates and quintessences that, in but an hour, change form, consistency, and color several times. Thus it is with blood, for the body's nourishment, and with sperm, which is the last operation of the vegetative soul or faculty. For sperm is, as it were, a masterpiece in nature, having the wherewithal to procreate its kind and, through these means, to perpetuate its species, rendering it immortal.[4] Thus, when one sees some lusty girl, more manly in manners and strength than her consorts or companions, one can well say that if she had been engendered an hour later, she would have been a boy; as, on the contrary, of a soft and effeminate boy, that one hour sooner, he would have been but a girl.

CHAPTER IX
WHETHER IT IS TRUE THAT AN OLD MAN CANNOT BEGET SONS

This proposition would be unworthy of refutation (since many women are seen bringing forth males even though their husbands are old) were it not for the suspicion one can have, and the doubt, that they are not really legitimate but are borrowed from a young friend. Thus, in order to safeguard and defend the honor of decent women who are often wrongly suspected of having some lusty man at their command to supplement the failings of the old husband, I am happy to oppose and overturn this false opinion, especially since ignorant people are convinced that an old man is totally unable to beget sons, to the point that, if things happen otherwise, they suspect there has been some borrowing.

It would profit me nothing to lay a foundation on the observation and experience of those who have had sons in later years (and whose wives have always had excellent reputations), not even if they were willing to try their wives by fire because of their conviction that they would not be burned, founded upon their firm belief that their wives have been chaste and loyal to them. For my opponents will doubt me, if they choose to, maintaining that the wives could have been so discreet, secretive, cunning, and sly that the breach of their wedding vows went unnoticed, and that they are held in high esteem when such is not the case. And as for my doubters, they might be happy to believe the wives chaste, but they would wish to know through vigorous reasoning how it is possible for an old man (who is normally cold, phlegmatic, and full of catarrh) to beget a son. For they are willing to allow that he can have daughters as long as he is able to beget.

I know perfectly well that there are enough mean and wicked women who, profaning holy matrimony, have no shame in deceiving their husbands, and that a smart woman never lacks heirs; for if her husband is impotent, she procures a pleasant companion who will furnish her a son to inherit the father's wealth under this latter's protection and nourishment, and if the father should die thereafter, everything will be from the mother. Now, I am not speaking for these strumpets. I wish to take the part only of decent women and to remove the blame, or the suspicion, that one could wrongly and without cause entertain concerning them.

I respond by saying that the old man can beget a son naturally in two ways that are fairly frequent. One way is when the woman's youth can correct and temper the sperm of the old man in such a way that it becomes able to form a male embryo, as we have shown in the fourth chapter.[1] Let us suppose that the woman's constitution is hot and dry, with a cleared womb and fine and bilious blood. From these conditions and circumstances, the sperm of the man will receive such an altering and tempering that there will be begotten from it a good male child. And who would doubt this? I even will grant that a woman getting along in years can nevertheless be of such a constitution that her womb will correct the cold sperm of her husband. I leave aside the fact that women, desirous of having children when they are kept from it because of some natural obstacle, use all types of thin-leaved mugwort and clary to warm up their wombs.[2]

The way I now come to is no less frequent: it is the old man's constitution, which can be hale and hearty, as is seen in some septuagenarians, and even in older ones, who do physical work with their arms and legs that another man of forty could not do. Why can he not be as vigorous in his genitals as in his other members? There are some who are stronger in some parts than in others: one is strong in the arms and weak in the legs; one is as strong as an ox in the head (even though he might not have horns),[3] another especially in the shoulders. Why would someone not also be stronger in the genitals than in the other members, in such a way that his greatest vigor resides therein?

Stop to think! Does one not see old men who are quite fiery and rigid, with little or no catarrh or phlegm, handsome and with good color? Why should it be that they not have a shot of hot and dry sperm to beget a son? Add to this, if you will (as I said of the women), that he uses warming remedies common to old men: spices, unwatered wine, and so on. I think it could happen on occasion that, with his wife well disposed for it, he will have the sperm needed for a male. Add to these reasons that the old man, wiser and more prudent than he was in his youth, practices this trade less often, now that the youthful fury has run its course and the nettles of the flesh have dissolved. He most often is contented with kissing his wife's nipples, tickling her belly, and indulging in other amorous caresses, delicacies, and entertainments. Besides this, the Christian calendar is observed in every detail: no joining on dog days, during months that do not have an *r* in them,[4] during dry weather or when it freezes, during the four quarters of the moon, during the whole of Lent and other days of fasting, feasts of great devotion such as the feast of Christmas, of our Lady, and of other virgins, apostles, and martyred saints; likewise on Fridays and on the Saturdays when meat is not eaten. So much so that there are hardly any good days for him (or for his wife, to speak more to the point) other than the Epiphany vigil, Thursdays, Mardi Gras, three or four days after Easter, and the feast of Saint Martin. Thus it is that the sperm, staying longer in its vesicles, is often more elaborated and refined in an old man than in a younger one. And in fact, one sees quite often in youth and in the first years of marriage that they only have daughters and that in later years they have sons. This is because when their irons were hotter, they never stopped beating them on their anvils and never did anything right. Since then, beating colder iron, they do more efficient work, and of a stronger temper. And so, one must not calumniate wives who give their husbands male children. But they must be careful about their honor; otherwise, if they but let men think they are attracted to them, they will be quickly seduced.

CHAPTER X
WHY IT IS SAID THAT A MAN CAN BEGET AS LONG AS HE CAN LIFT A QUARTER-*SEPTIER* OF STRAW;[1] AND WHETHER IT IS TRUE THAT THOSE WHO HAVE DEEP-SET EYES WERE BEGOTTEN BY AN OLD MAN

This popular saying serves as confirmation of the preceding one, when the populace admits that a man can beget, even though he is old, as long as he can lift off the ground, without help from another, a quarter of a *septier* of straw. This is a very lightweight material, so much so that one does not need a lot of strength to lift it. Others say a *glech* of wheat, which means in Gascon a sheaf, thrashed and without grain, something lighter yet. What is meant by this comparison is that a very old man is able to beget; and consequently, his wife shall be considered chaste when she gives him children.

Aristotle in his *Politics*[2] estimates that a man can beget until he is seventy and that a woman can conceive until fifty. This is the common and ordinary limit, for one sometimes sees a woman go beyond these terms, which can be limited only by her flowers. Still, Elizabeth, the mother of Saint John the Baptist, conceived when she no longer had her flowers.[3] But this was a miracle, as we shall say in Book Three.[4] According to nature, a woman can conceive only as long as she has her natural purgations, which seldom continue beyond fifty. Thus it is said that in Avignon, the wife of a tailor named André (a servant of Monseigneur Joyeuse), a woman of a very strong constitution, continued having children until she was seventy. Similarly, there have been men who at seventy-five years and older have had children, without any suspicion that they might not be the fathers.

But there may well be imposture or abuse, as in the case of a rumor that has been circulating for about five or six months of a woman from Saint-Denis (nine miles from Carcassonne)[5] who, very decrepit and claiming to be about eighty, said she had her last child at the age of sixty-eight and is today (which is February 27, 1578) pregnant again. It is also said that her husband, who died two months ago, leaving her with child, was over a hundred. Madame de Joyeuse, a true lady and full of every virtue, sent to have the truth of the matter ascertained. It was found that this woman, a poor beggar, is mistaken, or that she wants to fool people in order to have visits as a supposedly miraculous case, seeking to receive goods and alms from those who visit her. She has been making such claims for over a year now, and the awaited child has still not been born. She can well have a swollen belly and feel movement in it, just as it happens in young women who think for several months they are pregnant, then their bellies return to size after letting out large quantities of water or of wind. It is much more believable that a man might beget at a hundred than that a woman might conceive at eighty. But in my opinion, the wife must still be quite young in order to conceive from a very old man who otherwise can be very healthy and of a sound constitution.[6]

For there are, in fact, men who at ninety are more lusty and vigorous than others at sixty. Some are older than a hundred in the mountains of Vivarais and of Dauphiné and in other harsh places, where the people have a very sober and toilsome existence, partly by custom, partly by necessity, living in the open air, with good water, millet bread, chestnuts, vegetables, bacon, and cheese for the most part. I have seen some who are one hundred twenty years old, as proved by their wedding contracts. Well? Is not someone who is to live to be over a hundred with the strength to keep working and walk without a cane not still vigorous at seventy? If he meets a trollop who is ready and willing, will he not be able to impregnate her, since he is still able to plow?

There is no exact limit one cannot go beyond, for years alone do not set the term, but the makeup of the body and its use, as with a piece of clothing that might be considered old when it is very worn, even if only three years old, but which has been put on and worn so much that it becomes threadbare and rough and tears easily. On the contrary, there might be a piece of clothing made twenty-five years ago (such as for a wedding) that could be considered brand-new because it has been well kept and looks new and not worn. Likewise, one can truly say that a man who is worn out, broken, and stooped is old when he is forty years old; and another, sixty years old, young and fresh when he appears strong and scarcely weakened. The passing years are not as important as the wearing down. This is, I think, the origin of the popular reply given when one asks a person's age: that years are lease periods for houses and chambermaids. It is useful to keep track of years for the payment of rent, but with people, years mean nothing in the consideration of one's state and present disposition, which are more or less what makes a person last.

Old age is, properly speaking, the wearing out of the body, brought about mainly by mental effort, troubles, and extensive undertakings, accompanied either by excessive exercise or by negligence of the body. The former breaks and crushes the body, the latter rots it. One sees courtiers quickly worn and aged from running after posts, from being too often in a standing position without moving from their places (which tires the legs considerably), from having their heads uncovered in midwinter, staying up late, eating on the run, having no fixed hours for meals, and doing other such hurried, urgent, and unreasonable tasks. Then there are the court jealousies, the yearnings, the concern for favor that rattles their brains with ambition, the avarice that gnaws at their hearts, the envy and hidden hatred, calumny, detraction, intrigue, and other court vices that eat out their insides. Who could live long and become old in such a captive and wretched existence? Also, those who live their lives sitting, like writers and financiers, become old (that is, worn-out) quickly for lack of exercise and because of wracking their brains. For out of inactivity the body rots, like a piece of clothing that is never aired out; and a continuously running mind saps the body.

The peasant, on the other hand, who always lives in the fresh air and works regularly, neither too much nor too little, and takes his rests at regular hours, always on the same schedule, will have an assured and calm mind free of violent emotions constantly working away at him. Such a man will preserve himself a long time and keep himself whole in body and in spirit. So much so that at sixty years of age, even seventy, he is stronger, quicker, and more nimble than a city dweller at forty. He will stand more pain, run faster, see without glasses, have all his teeth, have a good appetite, and digest richer foods. He will not be gout stricken, full of catarrh, nor otherwise subject to illness. And who could doubt that such a man will not still father children for a long time to come?

In conclusion, I will say that a man's strength, both in matters of begetting and other actions, cannot be limited according to precise ages, which are nothing more than arbitrary numbers, but is rather a matter of constitution and of one's state of health, which sometimes stay sound for a very long time.

As to the other subject in the chapter title (that those with deep-set eyes have them because they were begotten by an old man), it is not resolved. For there are many who do not.[7] It is quite true that if the father was old, not only in years but in constitution and unhealthiness, his sperm was less vital. Thus, the body of the child also will be more slight and sickly because of it. One of the best indications we have concerning strength and healthiness is commonly read in the eyes, which change quickly in response to different dispositions, because they are so delicate and sensitive. Therefore, much attention is paid to them in illnesses, even for life or death judgments. Those who are wasted away and are low in natural moisture, such as people with hectic fever, have deep-set eyes because of their great dryness. In several animals that we eat, in young goats, for example, health is judged solely from their eyes. And so, it is likely that a child begotten by an old man will have eyes that are more deep-set, just as his whole body will be more frail and sickly, if such a man were his father.

But, as I have said, there is a considerable number of old men who are vigorous, strong, full of sap, and abounding in natural moisture. And there are many with deep-set eyes who are nevertheless very healthy, full of good, rich, and heavy moisture, yet who are known not to have been begotten by old parents. Their deep-set eyes have indeed another cause, which I am keeping secret here and discussing only in our schools on the basis of what Galen said in his book entitled *Little Art*, or *Medical Art*.[8]

CHAPTER XI
THE FALLACY OF WOMEN WHO DOUCHE THEMSELVES IN ORDER TO GET WITH CHILD; AND CONCERNING THOSE WHO WITH THE HELP OF FIVE HUNDRED DIFFERENT REMEDIES STILL CANNOT MANAGE TO GET WITH CHILD

Ignorant laymen are of the opinion that women are only sterile for one reason: a cold womb. So in order to become pregnant they often douche themselves repeatedly with certain concoctions of every kind of heat-giving plant they can find. These are for the most part Saint John's herbs, and women also gird their loins on this particular saint's day with these herbs, believed to have the property of rendering or keeping them fecund, even if put on the outside of their clothes.[1]

Now, the fallacy of thus washing oneself is most ludicrous, especially since all these women are not sterile because of coldness or superfluous moisture in their wombs, which may be a reason for the sperm's not remaining there. But very often it is just the opposite: their womb is too hot and burns or roasts the sperm or else dissipates, dissolves, and consumes the finer and more vaporous substance that is the essential part of sperm. This is why it remains feeble and withered, unable to fashion the embryo, and is soon expelled.

This condition is very common in women of a lusty and lascivious inclination, insatiable abysses for sperm, who are said to be hot as bitches and (if it were not lacking in respect) would run after men and take them by force, so chafed are their harnesses (which make them itch continuously and which are often as taut as a male's member).[2] Such "burnt whores" (as is said in Languedoc) have no desire whatsoever to get with child. They would need a pint of sperm each time, just to put out or calm down their fire and slake their thirsty wombs. For the little squirts that a man can give only serve to make their wombs hotter, as does a little water in a coal furnace, and to make them more and more thirsty, like the fever striken, who only drink a swallow at a time, which is why they keep starting all over. If hot douches are ordered for such gulping and absorbent sperm pits (whose cause is this great, voracious, and insatiable burning), is it not throwing oil on fire, making them run up and down the streets wild with such a thirst, in danger of throwing themselves down a well?

It is therefore necessary to be able to discern and to distinguish among the causes of sterility in women, so as not to worsen their condition, which requires contrary remedies to temper their wombs. They are most often too cold and extinguish the sperm; sometimes they are too moist, which also quenches it, drowning and rejecting it soon thereafter. At other times they are dry and arid, like sandy soil, lacking in moisture and therefore sterile. At yet other times they are hot and burning, which roasts and grills the sperm, so that it is unable to spread out, cover, and stick to their insides.

A womb that is cold and moist requires those douches women usually give themselves. The dry womb is upset by these baths, and even more so the womb that is too hot, in which case it is fitting to cool and moisten it, and not to heat it up all the more, as do ignorant laymen, following no method whatsoever.

One must also ascertain that it is not because of the husband, for one would treat the woman in vain, and all the douches in the world (both natural and artificial) would be for naught. Here is where many women are very often mistaken: they take the entire fault upon themselves, as if every man were capable of begetting, and as if it were only the fault of the woman.[3] It is like blaming the soil for everything when it does not make the seeds that were sown in it germinate. Can it not be that it is not the fault of the soil (which is in itself good, well cultivated, sown, and watered), but rather that of the seed, grain, or fruit, which is feeble, withered, rotten, or too old? So, too, the womb can be well disposed and the woman able to conceive, but nothing worth anything is put in her; or if it is good, it does not match her condition but would be better for another's.

Just as, in the same way, several seeds and fruits do not germinate or fructify in every soil, even though the seed is perfect and the soil very good, but they are incompatible, or there is insufficient sun, or the air is too cold.

Also, there are diverse hindrances, sometimes on the husband's part, sometimes on the wife's. Several women could conceive from another husband, several husbands could beget with another woman. And yet people always expect it to be the woman's fault that there are no children, unless the husband is old.[4]

Because of this belief, there are women extremely desirous of conceiving who use all the recipes in the world, rational and empirical, one after another without ever stopping. In this they are gravely mistaken and very often ruin their own fertility, which is not blocked but only slow in bringing forth children. They do not have the patience to wait for their natural time, but want within one or two years of their marriage to have children as they see the others do. And is it not well known that there are as many different constitutions as there are faces?[5] Animals and trees generally bring forth fruit more often than men; nevertheless, there are animals who reproduce only after four years, others only after six, ten, twelve, etc. Among trees also, some reproduce after the first year, others much later; it is said that the palm tree only bears fruit when it is a hundred years old.[6] Whoever would try to get plants and animals to speed up their times appointed by nature would gain nothing. The same goes for men, who have as much diversity among them as there is among all the other animals combined, as I shall amply demonstrate in Book Three.[7]

Thus, very often women trouble themselves in vain by drugging their bodies. What is worse, it sometimes happens that they mix up the cards to such an extent that even when the time foreordained by nature does come, they are unable to conceive. This is because at that moment they are not in the condition they should be in order to conceive. There is also another mistake: they mix so many recipes that one cancels out the other; and if by chance there happens to be one that is good, they cannot wait for it to work its effect, moving on to still another one if they fail to become pregnant immediately. Their poor bodies are so altered and mixed up by a chaos of remedies, and their minds so tossed about by hope and despair, that the sperm cannot find a safe port to its liking.

CHAPTER XII
WHETHER OR NOT A CONFIRMED LEPER OR SYPHILITIC IS ABLE TO BEGET HEALTHY CHILDREN

Some have doubts about this question; others believe firmly that the children of lepers and syphilitics are inevitably likewise afflicted. The truth of the matter is of great importance to the state and to the economy, for the company of those who are thus tainted by their parents must be suspect, and their education and regimen must be more refined and severely dealt with than that of children born of healthy parents. As in all hereditary illnesses (epilepsy, phthisis, or ulceration of the lung, nephritis, gout, and other such diseases), care must be taken for the children, who must be made to live under a stable regimen ordered by the physician, so that natural conditions and susceptibilities will not be actualized, or if so, only to a slight degree, and will be crushed and stamped out in the first generation without being passed on to grandchildren and great-grandchildren, as they do if, in the first and second generation, their condition has not been treated.[1]

Now, as for the two parts of the proposed question, I have answered the first (that concerning the confirmed leper) in the last problem of the third part of my treatise on harquebus wounds. After debating the question *pro* and *contra*, I concluded (as the popular saying goes) that the mortar always smells, be it a lot or just a little, of garlic.[2] This is why the company of lepers is dangerous. And it is (in my opinion) why people commonly say that leprosy penetrates seven fathoms of solid wall, that is to say, it is extremely tenacious in the areas in which it has taken root and spreads like strangletare.[3] This is why others affirm also that leprosy is caught by walking where a leper has merely passed by.

There, in a few words, is what seems to me to be the matter of the first question.[4] As for the second, which is about the syphilitic, it is not nearly as important a concern because syphilis is a less serious disease than leprosy, and also because it is a foreign disease, which becomes weakened little by little, so that in the end it will disappear entirely (as I shall prove elsewhere),[5] or it is nothing but a simple mange, which is also a contagious disease. For the moment, syphilis is as curable as several other diseases, which cannot be said of leprosy, which is absolutely and completely incurable, if it is confirmed. If, therefore, syphilis is curable, and some people recover completely, certainly children conceived shortly after the recovery of the father and the mother will not feel its effects. But the parents must be wholly recovered, as they can well be if they are of a sound constitution, have only barely caught the disease, and were skillfully, prudently, and diligently treated, as we shall point out in the sixth chapter of the twenty-first book.[6] Such people, once they are cured, will henceforth have sperm as pure and clear as before.

This is most certain, but it seems to me that what is being asked is whether men who beget or women who conceive while they still have syphilis, or are not wholly recovered from it, can engender children who are healthy. I will tell you: there are syphilitics who do not have a serious case; others have the disease more externally because of their robust constitution, which drives all the ill effects of the disease far from the principal parts affected—thus, the arms and legs suffer from a few spots or sores. If the disease is more external, it is possible that the sperm will not be infected by it, as is the case when the disease is hidden and deep and is said to have penetrated the very marrow. Furthermore, if the influence of the foul syphilitic property in the sperm of the father is light, it can be extinguished in the womb because of the good temper that the mother gives to it, softening it with her own sperm and with copious blood, which can predominate over this foul property, annihilating it completely. This is why the woman is often untouched by the syphilis that her husband communicates to her, because she is not susceptible to it and resists the disease, which her sound constitution masters. And so it is possible that a syphilitic man (not at twenty-four carats[7] and considerably weakened, but with good credit) might beget children who are healthy or at least do not have syphilis. Since they can be sickly and puny in other ways, they would, in common parlance, be called unhealthy.

CHAPTER XIII
WHETHER IT IS POSSIBLE FOR A WOMAN TO POISON A MAN THROUGH THE VENEREAL ACT[1]

It is written that a king of India sent to Alexander the Great a woman of exquisite beauty who had been fed poison from childhood, and she was thus so venomous that those who slept with her were poisoned to death. Aristotle, Alexander's preceptor, saved him from this treachery and foul deed by convincing him to abstain from the woman, whom he recognized as poisonous by her serpentlike, flamboyant, and wild eyes. He was also suspicious because the Indians practiced that kind of villainy. This was verified and proved at the expense of a few minions of the court who refused to heed Aristotle's warning: they enjoyed the woman carnally and died soon afterward, just like victims of poisoning.

Some treat this story as a fable, although serious authors such as Avicenna[2] recount it, calling upon the authority of Rufus.[3] Many others confirm the story, proving by natural arguments that this is feasible: that is, from childhood a person is given poison, little by little, as Galen recounts of an old woman of Athens, who ate hemlock as if it were any other food.[4] For, as he says, she first accustomed herself to eating small quantities of it, then a little more, and every day she increased the amount until she was able to eat a lot of it without harm. As for the Indian woman, it is said that she had been given, and accustomed to, *napel*, which means (according to some) "hemlock" in Arabic. But some class *napel* among the aconites, which are of a contrary nature: namely, hot, dry, and corrosive, not cold and benumbing, as is hemlock.[5] It seems that the wild, shining, and staring eyes of the woman argue sufficiently for a compound that was not a cold poison.

But Galen will not admit that a hot and dry, corrosive and burning poison could ever be converted into food, even if one took the minutest quantity; such a conversion can, however, happen with cold poison. This is because (he says) cold poison does not cause death by its nature but by its amount.[6] Such poisons are poppy, hyoscyamine, hemlock, and mandrake.[7] This is why it is so very difficult to believe what is written concerning this woman, and that she could poison others even with her breath, as it is said of the basilisk, which poisons on the outside those who but see it and hear it hiss, as Galen recounts, and that even when it is dead, every animal that touches it dies.[8] Other snakes are not exempt from this fate, even if they smell its breath, as Aelian says.[9]

I really think these are hyperbolic and excessive accounts meant to convey the idea of a virulent poison. For the viper, which is highly poisonous, does not harm unless it infects with its saliva some area where the skin is broken. But let us leave aside this woman and other such hardly believable artifices. Let us see if a woman could poison a man with the venom hidden in her, and which the virile member would either absorb into itself or pass along to the man's body.

I have never heard of this happening, but some claim it is feasible. Let us, therefore, examine the means, not of teaching this (God forbid), but of discharging this error and of acquitting the women some would wish to accuse of such wickedness, if possible (as I think will be the case at first blush). For in what part of the passage into the womb would she lodge the poison?[10] If she puts it deep inside, the poison will not come out when she wishes, in which case she will be in graver danger than the man, especially since the poison residing in all its strength and proper quantity in the womb will harm her much more than the man, who will only get its vapor or a small portion of it. I grant that the womb, because of its thickness and compactness, will resist poison seeping to the heart and other noble parts. But it will also become inflamed and, consequently, so will the entire body, as when a man poisons a woman in this passageway. If she puts the poison inside the neck of the womb, a similar problem will arise. And if she does not push it so far in but only rubs it on the inside of the large canal, the poison will not infect the virile member very much, not any more than it would by rubbing one's finger with some poison for a bit; and during this time the woman will suffer the greatest harm from it because of all the contact the poison is able to have on her insides.

Something else puts the woman in graver danger than the man: the man is not always ready for love's games, as is the woman. Hence, we answer the question of why it is that women do not seek out and beg men for the venereal act (in which they take as much, if not more, pleasure) but wait to be asked. Since the man is not always ready, as is the woman, it is a matter of his taking it malapropos, for (as the proverb goes) often the oven is hot but the dough is not risen. So, too, if the woman had prepared herself to communicate the poison (hidden in the passage to her womb) to a man who is to make love to her, it could well happen that he might not execute his duty as soon as she might have planned, and she will end up poisoning herself. For the poison could have had more of a delay and a stay than the woman had thought. Thus, such a means does not work for her.

There remains the placing of poison in her womb by means of some venomous perfume, as can be done with gloves, shirts, and other pieces of clothing: even saddles, stirrups, boots, and other equipment (for there is nothing one cannot soak with poison, whence the custom of kissing everything that is handed to a prince). I cannot understand how this is feasible without harming the woman's parts. For the vapors and perfumes absorbed by the womb are certainly transmitted to the other organs: witness uterine suffocation, which is caused by what is retained, against nature, in the womb. And when something that smells strong is applied in that area, such as garlic, or if perfume is used there, it is often smelled on the woman's breath. This is why she would thoroughly poison herself, given the insidious penetration of the fumes of the poison that one wishes to apply only to the womb. Now, it is one thing to treat gloves, clothing, and harnesses—inanimate things—and quite another to treat a living organ, which has a liaison and dealings with other organs and is not separate from them.

One can poison gloves secretly, and even a piece of food that will be on a platter, as Agrippina did with her husband, the Emperor Claudius, with the help of Locusta,[11] the famous poison maker, who prepared a beautiful mushroom, which she placed among others, from which Agrippina ate without danger, and thereafter Claudius had his little mouthful, from which he died the next night. One cannot, however, thus separately poison the living womb in the human body. For it has connections with the brain by means of the nerves, with the heart through the arteries, and with the liver through its veins. How would it be possible for this part alone to be soaked with poison and the others not to feel it in the slightest? Indeed, I cannot understand how this could be done, and consequently, I do not believe any of it. It would be absolutely necessary for the whole woman to be venomous (as is written of the Indian woman) or at least for her womb to be, and for this to be a natural, primitive, or acquired habit, so that this venomous quality could only harm others and not herself. Then her poison would be like the saliva of a snake, which is not poisonous to itself, but to us. It would be the same way with women if it were true (as they say) that her menstrual blood was poisonous.

For her menstrual blood does not poison her (even though it does her great harm in being retained against nature), as it poisons plants (by making them die) and dogs (which, as is said, become rabid because of it). If this were true (which it is not, as I have sufficiently proved in my *Paradoxes*),[12] I would verily admit that women were venomous and could poison men through copulation when they had their flowers. As regards giving a man some menstrual blood to take by mouth (as does a foolish woman in order to throw a spell on a man and make him fall in love with her, putting menstrual blood on the level of philters), this is a villainous practice, and no less foolish than indecent, just as is the one of using (for the same ends) a woman's navel, as I point out in the fourth chapter of Book Four.

I know there was a great legal proceeding against a woman who tried to give her husband some of her menstrual blood to draw him back from a wild fling. She was accused of trying to poison him, for ignorant people believe such blood poisonous. But they speak foolishly, since we indeed have need of it; otherwise, we would have been poisoned in the bellies of our mothers. Hence, one could also reply that, for the same reason, women's menstrual blood is not poisonous for us, as it is for plants and dogs. But these are stories: another kind of blood poured on plants will make them wilt just as much; and that dogs get rabid from it, I do not believe in the slightest.

And so I would naturally conclude that a woman cannot, without suffering on her part (which will not be the lesser), poison a man with the passage to her womb, even though he can catch leprosy and syphilis from her corrupt matter, as we said in the preceding chapter. And it is indeed poison enough to render a man miserable for the rest of his life, if he is not treated early on. But to speak of poison is strictly another matter and circumstance, for it can hardly be enclosed in a part of the body without causing it great damage and without diffusing throughout.

One will say to me that there are slow and insidious poisons (that is to say, long-term) that will not harm a woman as quickly as they will a man, and so she will have more time for an antidote, especially since she knows what is happening, while the man is unaware because it is being done through treachery. And the poison will be felt sooner by the man because he is hotter by nature, and the poison is hot. The woman, because of her natural coldness, will resist more. Likewise, the virile member will absorb it more readily through its wide and dilated opening than will the womb.

These are possible objections, but I am not moved by them enough to change my first opinion. For it is certain, from the conclusions I have drawn above, that a woman could in no way be exempt from harm in doing this. Or else it would be necessary for her to have a little poisoned tube or gut inside of her which she inserted into the passageway to her womb and lodged it in such a way (with the wide end toward the outside) that the man would put his member inside of it. This could be done by those who have one big enough in every dimension. And it would be necessary for the tube to be soaked and softened with fat or some other liquid, or for it to be an animal's intestine to provide smooth passage for the member. But who would be so dull and witless as to get caught in such a snare, or who would not notice it immediately and pull out of such a quagmire? Moreover, he would be too forgiving if he did not kill such a villain on the spot, at least if he were strong enough.

I would rather hold that all this is impossible, and that the woman could not manage to do this without the man's noticing it immediately, unless he were very drunk or completely witless, like a blind man who, after much merrymaking, puts his thing into the leather pouch of a boy dressed up as a girl to play a joke on him, and like one who, making love to his wife, was putting it somewhere under her fluffy petticoat when she said to him: "Monsieur, it's not where it's supposed to go." "That doesn't matter, my sweet," he replied. "It's in a warm place."

END OF THE SECOND BOOK

THE THIRD BOOK OF POPULAR ERRORS CONCERNING PREGNANCY

CHAPTER I
HOW IT IS THAT A WOMAN CAN HAVE NINE CHILDREN IN ONE BURDEN

In Agenais there is an illustrious house by the name of Beauville, once very opulent and great in wealth and reputation, from which came the most virtuous Madame de Beauville, today the wife of the heroic, valiant, and hardy captain known throughout the world, Sire Blaise de Montluc, most worthy and meritorious marshal of France.[1] It is said that the grandmother of Madame de Beauville had nine daughters in the same burden, all of whom eventually married and had children. The mother and daughters were buried, successively, at Saint Crepasi, a school chapel in Agen, founded and built by the house of Beauville. The mother had her sepulcher raised on a portal among nine others fashioned for her daughters in memory of the event.

I saw a few of them recently, having been in Agen in 1577, in this very chapel. The others had been destroyed during the civil wars.[2]

Here is the story: Madame de Beauville had a beautiful and healthy servant girl with whom the lady's husband seemed enamored. In order to get rid of her in a decent manner, she married her off. This servant girl had three children in her first pregnancy, and the lady took the notion that her husband had played a part in it, unable to convince herself that a woman could conceive such a number of children from just one man. So she became doubly jealous and, no matter what was said to the contrary, began to hate the poor girl all the more.

Then it happened that the lady herself later became pregnant and grew so large that she brought forth nine daughters. This was interpreted as divine punishment for her calumny: she would be accused of the even greater sin of having slept with several men (for she always had maintained stubbornly that from one man a woman could conceive only two children because a man had but two stones of generation). And so, greatly shamed and fearing defamation and condemnation by her own pronouncement, she was sorely tempted by the evil spirit (who had led her to such despair) to have eight daughters drowned and to keep but one. She kept the matter secret, between her midwife and a chambermaid; to the latter she delegated the horrible task.

But God, who preserved little Moses from a similar misfortune, willed that the husband, returning from the hunt, should happen upon the chambermaid. Whereupon, discovering the matter, he saved his innocent daughters from death and, unbeknownst to the mother, had them raised and at baptism gave them all the same name: Bourgue, just as the ninth was named (the one kept by the mother). When they were a little more grown, he had them brought into his house, all dressed in the same material and in similar fashion, along with the ninth, who was already living there.

Having assembled them in the same room, the husband then called his wife, accompanied by relatives they had in common and close friends. He told her to call Bourgue, which she did. Upon hearing their name, each of the nine answered. At this the mother was most astonished, and seeing them identical in size, countenance, voice, and dress, she became dumbfounded. Suddenly her heart told her these were her nine daughters, and that God had preserved the other eight she had exposed and thought dead. The husband reproached her for her inhumanity before all who were gathered there, pointing out that this could well have come about to confound her for the mean opinion she had of him regarding the servant girl.

This is more or less how the story is told. Almost identical to it is that of the Piglets[3] of the city of Arles in Provence, the provenance of the house of Malliane. They were thus named because the chambermaid, who was carrying the eight infants to drown them, was met by the husband. She said they were piglets she was going to drown because the sow was unable to nourish so many. In memory of this they were named Piglets and have a sow in their coat of arms. They say that it was because of a curse of a poor woman who, surrounded by her small children, asked alms from a lady of the house in question. The lady reproached her for being lascivious and too much given over to men. Whereupon the poor woman, a decent soul, uttered the curse (as they call it) that this lady would get with as many children as a sow has piglets. This is what happened through the will of God, to show the noble lady that one must not impute to vice that which is a great blessing.

The same is said of the magnificent wittiness of the Scrova of Padua who bear in their coat of arms a sow, called *scrova*, surname of this family, a corrupted form of the Italian *scrofa*. One also reads in the annals of Lombardy that in the time of Algemont, first king of the Lombards, a whore gave birth to seven sons, one of whom succeeded Algemont. And Pico della Mirandola[4] writes in his commentaries on the *Second Hymn* that in Italy a German woman gave birth to twenty children in just two pregnancies; in the first she had eleven, and her belly was so big she carried it with a tablecloth tied behind her neck. Albucasis,[5] the great Arab doctor and surgeon, witnessed a woman who had seven children at once, and another who aborted fifteen that were well formed. Pliny mentions one who aborted twelve.[6] Marcin Kromer in his *History of Poland* writes that the wife of the Count Virboflaë in Cracow had in one burden thirty-six live children in the year 1269.[7]

Thus, several accounts bear witness to the fact that women can bear an abnormally large number of children in a pregnancy. Let us see how this can come about. I always put aside pure miracles, for unless one maintains this to be completely miraculous, I will not grant such a large number, and still less that of three hundred sixty-three, as is written of the Lady Margaret, countess of Holland, in the year 1313, during the reign in France of Philippe le Bel, as it is recounted in the *Sea of Histories*, in the second volume, in the chronicle of the Emperor Henry.[8] It is said that it was because this lady made fun of those who have more than one child at a time, affirming stubbornly that it was impossible for a woman to have two children at once by the same father. As a punishment for the words of this accusing calumniator of nature, she conceived and brought forth in the same burden three hundred sixty-three live children, like little chicks, who were all baptized. If this is true, it is a pure miracle, exceeding nature's limits (unless this lady was a giantess), and in miracles no cause is necessary other than the pure will of God. For

He is all-powerful and, making everything from nothing, will (if He wishes) make every hair on our heads become a child, or from every pore and hole in our skin exit a full grown man, just as big, fat lice come out of those who have the disease called phthiriasis in Greek and pediculosis in Latin.[9]

In this matter of miracles one must not be stopped by the capacity of the space nor muse upon the sperm, or any other substance. Nothing is impossible with God, sole author of true miracles. But since He only does them for a great mystery and so that they might be held in much reverence, He wishes them to be very rare. Just because one sees something strange and prodigious, it is not to be taken as a miracle, such as the abstinence of a few people who have gone two or three years and more without drinking or eating, for a natural reason which I have sufficiently explained in my *Paradoxes*, where I excepted the fasts of Moses, Elias, and Jesus Christ as truly miraculous.[10] In the same category are the miraculous pregnancies of the Virgin Mary and of the holy women beyond the age of childbearing, according to the laws of nature, who were sterile, like Sarah, Abraham's wife, who was already ninety years old (for which reason her son, Isaac, is called child of promise and of spirit),[11] and Elizabeth, mother of John the Baptist, whose age was an argument in convincing the Virgin Mary of the mystery of the incarnation of our Lord Jesus Christ: "And, behold, thy cousin Elizabeth, she also hath conceived a son in her old age."[12] This expressly signifies a miraculous conception, and that nothing is impossible with God, who changes and alters as He pleases the order of nature He has established.

If one maintains that these burdens of nine children are pure miracles, it must no longer be discussed but simply believed.[13] Because we are not bound to it, since it is not in scriptures nor approved by certain holy personages, we will be permitted to inquire through reason if this can take place naturally, and by what means. We notice every day that there are very strange and rare things that come about through natural means, which are themselves also rare. We call them natural miracles, or miracles in nature, as distinct from supernatural and divine miracles, in which nature has no part and for which there is no foundation in nature.

Natural miracles are (if you will) things such as females having children at nine years of age and men and women living two or three years without drinking and without eating. Or a mule having a colt, as we saw in Montpellier last year (which was 1576): it was a large plow mule that had been brought from Agel near Béziers, and which was still feeding with its milk its big and beautiful colt. Or a woman carrying in her belly for more than six years two dead children, or at least their bones (the soft parts having been dissolved and expelled in the form of pus), rejecting little by little in great pain all the bones, dry and white, from which it was known that they had been twins. Two or three years later she became pregnant with two more, which she aborted in the sixth month while strenuously walking around in the bedroom: the arm of one of them dropped out by itself onto the floor, then the rest of the dead body (not rotted) came out, which after about fifteen hours was followed by the other one, completely whole and dead also. This we know for a fact to have happened also to a virtuous lady from Frontignan, twelve miles from Montpellier,[14] married to Jacques Galhard, a rich bourgeois. Mathias Cornax, a physician from Vienna in Austria, tells of a woman who carried a dead child in her womb for more than four years, which was finally removed by an incision made in the belly; one year afterward she became pregnant with another son. Likewise, another carried for thirteen years all the bones of a child in her belly, and a third who after a year pulled the bones of a dead child out of an aposteme on her belly. I leave aside a number of prodigious and monstrous natural things I have in my small rooms that come about against the order of nature, and which I will most willingly show. Hence, one can be convinced that other cases, as unusual or even more so, can occur. Let us now consider, I beg you, how this can be.

Animals usually have twofold wombs, like two horns, and each horn has several divisions, like cells or little chambers in which the little ones are separately lodged. And there are naturally as many little booths as the female has teats, and she is thus able to nourish as many as can be conceived through nature's providence. A woman has only two mammae and can thus carry only two children ordinarily and nourish the same number. For if she has three or four at a time (as we have seen with one woman in Aubenas in Vivarais who had two children with her first pregnancy, three with her second, and four with her third),[15] one will harm or hinder the other, and they will not live communally; ill nourished in their mother's belly, and thus unable even to endure the strain of coming out, they die in the birth passage, or soon afterward.

Yet in Aurillac in Auvergne, the wife of a man named Sabatier brought forth three sons in one burden: the first and the last lived twenty-four hours; the one in between (who because of this retained the name John of Three) grew up and was married in Paris. He died not long ago. Similarly, Master Ambroise Paré,[16] first surgeon of the king, most learned, curious, diligent, and frank in publishing the treasures of great wisdom and experience God bestowed upon him, observes in his book on monsters that in Sceaux (between Chartres and Maine) the Demoiselle of Maldemere had two children in the first year of her marriage, three in the second, four in the third, five in the fourth, and six in the fifth; from this last burden came the Sire of Maldemere, still living today.

Aristotle affirms that in Egypt it is not rare for a woman to have five, and that a woman was seen there to have had in four pregnancies twenty children, five in each one, and most of them grew to maturity.[17] Aulus Gellius witnesses in the time of Augustus Caesar that one of his country chambermaids had five children at once, none of which lived long, nor the mother after them.[18] Master Guy de Chauliac, toward the end of the second doctrine of the fifth treatise, mentions five, seven, and nine, based on the authority of Avicenna and of Albucasis. This is the largest number the ancients record, exceeding by far the number of a woman's mammae, which naturally correspond to the number of the burden.

Regarding the womb, it is not divided in two, in the manner of animals, and does not have booths separated one from the others, as some, ignorant of anatomy, have imagined and then written according to their fantasies, saying that there are three chambers in the right horn, where males are formed, and as many in the left, for females, and one in the middle where, on a few occasions, hermaphrodites are begotten, also called androgynes and, commonly, women-people, who are of both sexes. These are ravings, as is everything said about such divisions and little chambers, for in truth the womb has but one cavity, as does the stomach and the bladder, and a child fills it up completely. If there are two children, each is able to have its bed, or secundine, which makes for the separation, and thus the woman is very big when the last months come. Sometimes both are conjoined in one secundine except for the child's-shirt, which is their wrapping, loose like a little skin, and which separates them.[19]

The beginning of this year (which is 1578) in Andule, a city in the Cévennes in Languedoc, the wife of a judge had three daughters on two different days (one on Friday and the other two on Sunday evening) and had but one secundine for all of them, which came out after the third child. They were only in their eighth month, were baptized, and lived seven or eight days.[20] Luigi Bonaccioli of Ferrara[21] declares in the third chapter of his first book on illnesses of women that one woman had one hundred fifty children (each with its secundine) as long and as big around as a finger, but they were not to live, which is one of our requirements. Such were the twins that my wife aborted effortlessly in the year 1575 (to my great regret and unhappiness) in about the fourth month. They were both in the same secundine and each had its child's-shirt. Otherwise, they would be joined, as though fused together like the double children we see, called monstrous.

But the mere child's-shirt, or biggin, separates them perfectly. If there are more than two of them, they can also be happy with one secundine, and the womb will contain them more easily and will nourish them better. For this secundine is very often as large, taking as much space and consuming as much food as the child itself, sometimes even more. Whence, one sees women so unusually huge that one thinks they will have twins, and yet they have only a very small baby, but a cumbersome afterbirth causing more trouble than the child.

And so I would maintain that women who have more than two children at a time do not have as much of an afterbirth because it is greatly reduced due to the space and food requirements of multiple offspring. I also believe that such children are borne only for a period of seven months, no less a viable limit than nine. This is because the womb had slightly stretched as much as several small but nonetheless viable babies would require. For there is no disadvantage to their being born famished, enfeebled, and wrinkled because of poor nourishment: it is enough that they are well formed and have all the parts needed by the nutritive faculty. They recover well from their fasting and abstinence if they find proper nurses who give them lots of milk. They make more progress in eight days than others who are born well nourished do in three weeks. Everyday we see very small ones born, all withered and wrinkled like an old apple, who in almost no time become marvelously big and fat. Even if the four or five from a burden were like small starvelings, provided they are healthy and have the strength to suckle, I have no doubt that they will be just fine, provided they are properly seen to.

And is it not possible for all these circumstances to converge in one pregnancy out of the five hundred million that occur in about a hundred years? But, some will say, it is a lot to have nine children and to have all of them live, or even five, as it is written of a woman from Bern, Doctor Gelinger's wife, who had five children in one burden; or seven, as did the bondwoman of a man from Siena, as told by Monsieur d'Alechamps, a most learned physician.

We must, then, in order to go beyond this belief, supply further corroboration for our arguments. And what more corroboration need we offer than to predicate that such women were of the largest size imaginable (which is very likely): big, tall, large in the flanks and hips, thick limbed, well quartered with heavy-columned thighs, low jointed, having a nice ample womb unencumbered by fat from the parts surrounding it, and as dilatable as one could wish. Also, the rest of their bodies would correspond to the lower parts: well supplied, juicy, and well nourished, neither hungry nor withered, so that there would be enough rich blood throughout the body of the mother to nourish several children at once. Can one not see that, with women of such corpulence, in one alone there are two or three teeny little women?[22] An arm larger than three or four other ones together; the same with the thighs, and all the rest proportionately? So much so that—can it not be said—one big beautiful woman is the equivalent of two or three teeny little women together? And if each of these teeny little women can have two or three children in one burden, as is seen often enough, even up to five males (as I heard of a small hunchback, the poor wife of a capper, in the city of Rouen in the year 1550), why could not this great woman have as many as the three she is replacing?

I do not mean that this is an ordinary occurrence, not any more than for teeny little women to have three or four. But I do maintain that it can happen, and that it is not more astounding for a large woman to have eight or nine than for a small one to have two or three, as long as she has a womb large enough and as much menstrual blood as three small women. Now we have our woman just as we wish, ready to conceive; we need but the male to provide the assignment and to put enough material for nine children into the oven to complement the woman's contribution. For she also has sperm that joins, allies, and unites itself to the man's, and it does not all go into the enclosing exterior, like the shell of an egg, as some think Aristotle's words imply; these people maintain that this exterior is the beginning, the basis, or foundation of the secundine. For if such were the case, children would not resemble their mother as often as their father. But the truth is that the woman contributes to the principal substance, from which the child's body is formed.[23]

Up, then! Let us see to it that women stay occupied, always ready to do their duty, ready to conceive and to supply a goodly quantity of their sperm, as though having amassed it and stocked it for a long period of time while their husbands were not with them. Here he is, coming from afar, traveling short distances every day so as not to arrive tired or spent, as do those who come posthaste in order to show themselves more affectionate to their other halves, but as soon as they are in bed they fall asleep. I would rather he return at his ease and arrive strong, healthy, refreshed, and joyful, as full of love for his wife as she is of desire for her husband.

I suppose that the male half of this Platonic androgyne will complement the corpulence of its other half, that he is large and well furnished in all parts and especially in the main one.[24] Let him not be fat and plump, for where there is much fat there is little sperm, nor angry and melancholic, for such people do not have much sperm either. I presume him to be jovial and of a loving nature, a lively and easygoing man. His spermatic vesicles, and the seed vessels on the end of them at the neck of the bladder, are about to burst for having been kept such a long time from love. Both being thus armed to the teeth and with munitions piled up, coming to an encounter to wrestle and to scuffle with the utmost longing, our couple will doubtless in the first round unleash a massive effusion of whitened blood, both on one side and the other. There will easily be enough for three or four children, since without such supplies others have just as many.

I would like for it to be morning when the gentleman arrives home, and for him to find his wife in bed. If he lays charge again in a little while, after having rested a little, they will have at it almost as violently as with the first tilt. And there you have the makings of another two or three; that adds up to seven, or what it takes to make them. Then they must have breakfast or the noon meal in a hurry. A short time before supper, when the company that had come to see them has left, they withdraw and begin kissing each other, and if there is no movement downstairs, they complete the task there or else find the remainder of the satisfaction in bed. For to put it off until the next morning would be too much to ask of people so greatly excited. This can be added to the other spillings, enough to make another one or two children, if not more. So there can easily be enough substance for them if the womb retains it well and conceives (as I always presume) enough to mold and to form ten children at once; but I will be satisfied with nine.

We can also add up our numbers differently, if we wish: suppose the husband arrives at suppertime and, having received the visit of his friends, retires rather late. This is the way things usually happen. If he is among the most vigorous in love's assaults and battles, he will lunge in and charge again and again five or six times all through the night. For he is rested up and has amassed a good supply of sperm, as we can imagine. This is not too much bed work for our good fellow, considering the story they tell of a young gallant who promises a lady he will make love to her twelve times in one night. He was well rested and ready. When it came time to mark the eighth of his twelve tilts (which he was doing with a piece of chalk on the bed board) the good lady felt compelled to point out that the last round was worthless. The young gallant maintained the opposite but said that he was a good merchant. After some gainsaying the gallant said: "You know I'm beginning to go bankrupt; let's just say that nothing has as yet been delivered." Erasing all his marks, he added: "I want to start all over." And so in the end he found he had ridden post twenty times. But there is too much exaggeration in all this, making the story fabled and unbelievable. For my argument I shall be satisfied if the husband does it a half-dozen times at the most. The first three will have the most weight and discharge and will well be worth a pair of children each, known as twins; the other three will be good for one child each. Or, if you prefer there to be only five tilts, let triplets be made from the first one, twins from the second and from the third, and from the following two, one child each. This way there will not be any unreasonableness, absurdity, or impossibility (as could be said of the aforementioned gallant, to whom is attributed much more vigor than the Florentine whose epitaph at the cloister of the Dome in Florence reads: *Messer Concia, per haver' chiavato una lira & un' oncia* (that makes seventeen times in merchant's reckoning) *hic jacet.*[25] Still, I would grant that everything I said about the good husband could come about in the space of three or four days.[26]

There is only one doubt remaining: whether or not the sperm that is expelled on several different occasions can be assembled and succeed in making a burden. For it is held that everything is done in one stroke and not over several. This is what remains to be explained and resolved. For as regards the quantity of sperm necessary to form the bodies of nine children, I find no problem, since the man (as I have predicated with the women) can be worth three others in corpulence and in necessary provisions. As for the several different times, the case is not unusual for such a small interval as I am speaking of, from morning to evening, or twenty-four hours, or three or four days, since Aristotle admits to superfetation after two or three months.[27] It is true that he does not consider viable those conceived a considerable time after the first conception, "but if the second is conceived soon afterward," he says, "it can be perfect and can be born with the first, as if they were twins, as is said to be the case in the fables of Hercules and Iphicles, and as was proved in an adultery that yielded one child resembling the father and the other her lover. In another case a woman, having conceived twins, conceived again some time thereafter. She brought forth twins in the course of time, along with another only five months along: this one died immediately, and the twins lived. Another woman delivered in the seventh month a child that died, and two months later two that lived, etc."[28]

Since this is the way things are, if one does not want to grant that the sperm expelled over three, four, five, or six times, however close upon one another they might be, can unite and join together, it is not unreasonable to recognize these diverse strokes as an equal number of conceptions making for just one burden, and to recognize the children resulting from them as able to come out at similar intervals, as is often seen with twins, one being born four or five days after the other, so that one could say that they were similarly conceived, on different days, and not both together. But because they are so close together they are considered twins. Moreover, not long ago, in

Agenais, there was a burden of three sets of twins, each born eight days apart. It is written that a woman of Alexandria seen in Rome at the time of Hadrian had five sons, the last of whom was born forty days after the four others, who were born together.

And stop to think! Our practitioners maintain that a vigorous and robust woman can continue having her flowers quite regularly while pregnant, that for this reason she can be superfeted long after her first conception, and that the child will come out perfect at the time of his maturity.[29] But this does not relate to the question, for the woman said to have borne nine children in a single burden brought them forth at the same time, at least within one day. It would therefore have been necessary for them to have been engendered within twenty-four hours. Granted, it is not unreasonable that one man should do in one day what another (or he himself at another time) does in two or three days. And then it would not be unreasonable that, of these nine children, a number of them might be seven months along, the rest nine, and they might come to term at the same time.

Thus, it seems that everything is in harmony, and it is no longer to be doubted (if it is feasible that in some way we are able to understand through reason) that the accounts presented, having been duly witnessed, are true. And if someone objects that, in order to make matters come about as I say, I lay conditions in such number that they could rarely all coexist, I respond that when effects are rare, so, too, are causes. It suffices to suppose that nothing is impossible, and that one merely requires a convergence of causes that, separately taken, could happen in nature. Only their meeting is, in this matter, extraordinary and what makes the case miraculous.

CHAPTER II
WHETHER A WOMAN CAN CARRY LONGER THAN NINE MONTHS; AND HOW THE END OF THE PREGNANCY IS TO BE RECKONED

One can be quite justifiably amazed at the fact that man, the most perfect of all the animals in the world (since the excellence of natural things consists in fixed number and order), has no fixed time for his reproduction or for his birth, even though the most wonderful of nature's works is the power to engender one's own kind. There is not a single animal that does not have a fixed time for mating and copulation; outside this time the venereal act does not occur. There is also not a single animal that, gravid with offspring, will receive the male, except for the mare, as witnessed by Aristotle.[1]

But experience is to the contrary, at least in our region (I do not know if in Greece the mares are of a different nature),[2] as I have been assured by Monsieur d'Yolet, steward of Madame la Princesse de Navarre and one of the great squires of France, owner of one of the best breeding farms in Auvergne, a careful observer of unusual things, and a trustworthy reporter of what he has observed. There is no animal known that does not have a fixed time to carry its offspring, and that does not, within a day or two of the allotted term, bring forth its little ones.

Woman alone is always receptive, or, as people say in Languedoc: "A woman and a capon are always in season."[3] During all of the four seasons, every month, every day, any time is good: during all quarters of the moon, all feast days and vigils. If dog days are alleged to be dangerous for men, women will answer that dog day nights are not forbidden. Then, when pregnant, they do not hold back because of it nor flee the male. She can be full to the point of gagging and very often will still take delight in it, even hunger for it, as if there were nothing in her belly at all.

But what is even more strange is that she has no fixed term for carrying her children, as do the other animals. For sometimes she delivers at seven months, usually at nine, sometimes at ten and at eleven; all of these terms are good and viable. For one must not speak here of miscarriages, which can happen in any month and at any hour. A few, wishing to explain this uncertainty regarding the various terms of carrying children, have said that it is because women have no fixed and proper time or season for conceiving. And why do they not have a proper season, or men either, for coming together? Because they are not stimulated only by nature to reproduce, but most often by voluptuousness and carnal pleasure. For this reason man is considered more bestial and less reasonable than animals. They add that man is often the cause of the acceleration and variability of delivery when he returns to the pregnant woman, where he does nothing but spoil the work, just as would one who digs up the earth after it has been sown and the seeds have germinated.

But that would more likely be the cause of miscarriages than of the various viable terms in the seventh, ninth, tenth, or eleventh month. For constant agitation can cause premature delivery and, at the very least, will not make it late. Thus, pregnant women who do not have relations with their husbands after conceiving usually carry for up to eleven months; those who have just a few, up to ten; those who have still more, up to nine; and those who have the most terminate in the seventh. Or, from another point of view, since the fruit or seed has already been fructified, if it is agitated and shaken it will lose time because it has to take root once again and become implanted again if it wishes to grow; hence, it will be slower to mature than if it had not been moved. Thus, the child who is shaken will be born much sooner; and the one whose mother is left to rest will be born later.

They also maintain that a poor regimen in pregnant women causes early or late deliveries: acidic, spicy, and hunger-inducing foods, fits of anger and other emotions, violent exercises and excessive movement in dancing, and other commotion of body and mind. But, again, these should be blamed for miscarriages and premature deliveries rather than for the variability of terms. Or it has to be that there is but one term set by nature, namely, the eleventh month, and that all the others are accelerated or early for the aforementioned reasons. And the question would still be unresolved: how can these other terms be viable if they are not ordered by nature? For it can also happen that an animal, because of some strain, will bring forth its young a few days or a few weeks early; but the little ones will not live. And children survive from four different terms: seven, nine, ten, and eleven months.

I now wish to turn to another topic, especially since I have not written this work in opposition to philosophers and physicians, people of my own profession, whose opinions and arguments I refute elsewhere when they seem false and absurd. Here I only have a quarrel with laymen, refuting their errors and instructing them in all modesty in what they wish to know. Thus, if they wish to learn what I think to be the cause of this diversity, I shall explain it to them simply, leaving to those who are wiser than I the final determination.

In the unique species of man there is as much diversity as in all the other species of animals combined,[4] which are infinite in diversity in quadrupeds, reptiles, aquatic animals, and birds, of which the individuals are all very much alike in every quality, scarcely differing one from the other except in size, based principally upon their age. Find me any other difference between two carp, crows, frogs, scorpions, or sheep, other than color or some small mark. Still, even these can spring from their species, if one will but pay close attention, and such differences distinguish their species from others, not a properly specific one, but an accidental one, as our logicians say.

But man, in his individuality, is so full of differences that no two will be found alike in the whole world; or if two are found, it is considered a great spectacle. Thus, I affirm that in the single species of man there are more differences than in all the other species of animals combined. I am not considering here other differences, which are infinite. I am only considering differences in complexion, whence spring all natural actions. We say that there are but nine complexions: one tempered and without any excesses, and the others, which are excessive in some simple quality (such as heat, cold, moisture, dryness) or double quality (such as heat and dryness, heat and moisture, cold and dryness, cold and moisture).

This is said generally, for every complexion is related to one of these; but each has huge differences in degree. That is, every hot complexion is not such to an identical degree, but one man is hot to one degree, another to two, a third to three. And these degrees are further subdivided, so that one is hot only to a half-degree, another to a third, still another to a fourth, another yet to an eighth, and yet another to a tenth, etc. And thus, too, with the other complexions, which are likewise of the cold, moist, or dry sort, insofar as these qualities are in excess. And such infinite differences proceed proportionately from diverse actions, not only natural and vital, but also animal, which are infinite in the species of man.

This diversity is not seen in any other type of animal. All cranes are of the same complexion, have the same behavior and the same activities, take and like the same food, make their nests in the same manner, etc. All domesticated oxen share the same condition, all the wild ones another. All the dolphins in the sea have the same temper, the same behavior, activities, and food. The ants under the earth are all alike, as are all honeybees; each species has its endeavors, its discipline, and its artifices, without a single ant or bee doing anything different from its consorts, because they are all of a single complexion, condition, and nature. Cicadas all have the same song, all cuckoo birds say cuckoo, and all birds have their particular jargon and feathers. All dogs bark in the same way, or very nearly so (the main difference is in the deepness of the voice);[5] so, too, with the lowing or bellowing of oxen, the bleating of sheep, the mewing of cats, the braying of asses, the neighing of horses, the cawing of crows, the croaking of frogs, the cooing of partridges, the cackling of quail, the peeping of chicks, the clucking of hens, the grunting of pigs, the roaring of lions, the howling of wolves, the yelping of foxes, the trumpeting of elephants.

But in the unique species of man, how many different voices there are, how many different languages, ways of singing, manners, ways of drinking, eating, sleeping, dancing, walking, running, fighting, arming oneself, riding, or traveling! How many kinds of trade and business, occupations, management, dealings, and enterprises! What diversity of conditions, emotions, and imaginings! It is infinite for the person who is careful to pay attention. To understand it more fully one has only to consider those who are in the same province:

what a difference there is between each one, not only in the cities they live in, but within the same city, or even the same house. One will want roast meat, another will want broth; one cold, another hot. One is quick to anger, another easygoing; one greedy, another wasteful; one lustful, another continent. One wants to be a monk, another a soldier; this one likes to be gorgeous, that one takes no care of his person; this one likes music, that one cooking; one hates wine by nature, another is always drunk. Moreover, some hate bread against every human instinct, others cheese, still others oil, eggs, fish, etc. Some faint from merely smelling apples or roses.

Why is that? It is because they are all of different complexions. Thus, also, some are hasty, others slow, some bubbly and lively, others lumpish and cold. Some naturally listen, others always want to talk; some are very friendly and very thoughtful, others hate everything, concern themselves about nothing, and are completely indifferent. Some are very interested in games, others do nothing but work. Some take up reading and become erudite, others wish to learn neither how to read nor write. There are some who are kind and sweet as angels, others are worse than devils. All this can be in the children of a same family, all from the same father and mother, raised in the same place. Notice, I ask you, what diversity there is in the same house because of different complexions, and judge from that how many there could be in an entire city, then in a kingdom, then in the entire world.

I now wish to accommodate the fruit of this discussion to solving the proposed question. Since the diversity of complexions is so great in man, and not among animals, one must not be surprised that man has no limited seasons as far as lovemaking is concerned, nor any fixed term for carrying children, as other animals do. As for the length of pregnancies, the diversity of the terms comes from the diversity of the complexions, both in the child conceived and in the mother. For there are children of great fatness and corpulence, who require more days for their maturity, as Aristotle says of elephants, which need to spend two years in the womb because of their corpulence.[6] Mares for the same reason carry for twelve months, and the ass likewise. I remember a Florentine matron who, upon delivering twelve months after her husband knew her, convinced him (as is written in the book of *Joyous Adventures*)[7] that, if a woman sees an ass on the day she conceives, she will carry the child as long as an ass does. To such a big fool as that (not at all in keeping with the norm of his nation), a putative child would be needed, whose term would be that of a stupid animal. Thus (to return to my point), a large fruit does not ripen as fast as a small one. This is why if one child, small and thin right from conception or first formation, hot and dry in complexion, moving and kicking, has enough with nine months (and sometimes with seven) for his maturity, another might need ten or eleven.

Thus, one often sees daughters come toward the end of the ninth month and sons born at the beginning and opening of this same month. Similarly, the mare carries a horse twelve months, but if she is gravid with a mule (as when she has been covered by an ass), she will carry it for thirteen months, as the aforementioned Monsieur d'Yolet has informed me. This is because mules are animals that are colder than horses. For a hot complexion hastens maturity, while a cold one matures later, as with fruit.

So much, then, for the child that, because of its complexion and the corpulence that proceeds from it, has a shorter or longer stay in the womb, awaiting ripeness. Cicero uses this term when he says in his book *On the Nature of the Gods*: "Diana is invoked in childbirth, since the child ripens in seven or in nine revolutions of the moon."[8] One speaks in such a manner, since the child is, properly, a fruit, made from seed; it ripens in the womb, as in a pod or a husk that opens when the fruit is ripe, ready to fall. This is what the womb does, which is so tight against the child during pregnancy, even toward the opening, that nothing can be inserted into it. When the child is well ripened, the womb opens as widely as the child requires.

Now, the swiftness or slowness of this maturing is not entirely the result of the child's complexion. The womb has an important part in it, even the principal one, to tell the truth. For according to its condition the fruit will sooner or later ripen. It is true that the pliability or recalcitrance of the fruit has much to do with it. Just like an oven that bakes bread: smaller and narrower loaves bake sooner, and at the same heat a partridge roasts sooner than a piece of beef. It is the fire alone that controls the diversity of the effect and the conditions of the various substances. Thus, the heat of the womb plays an important part in the rapid or slow maturing of the child, which, by the way, has within itself the wherewithal to mature; and this is how it differs from the bread and meat to which we compared it.

The same could be said of the sun and the fruit it ripens. Fruit has within itself a natural heat that brings it to ripeness, but the sun, hitting it, speeds things up much more. Thus, we see fruit on a tree ripen at widely different times, one today, another tomorrow,[9] and so on for a month, now one, now another, but not all at once; there are different degrees of ripeness. So it does not all fall at once, if left alone, because it has not finished ripening. On the side where the sun hits it, it ripens sooner, and since the sun in its natural movement turns around the tree higher today, and tomorrow a degree lower, all the fruit thus ripens.

The womb and the entire body of the mother act likewise with respect to the child. Thus, one must not find it strange if, of the two twins conceived together, one is born four days before the other. For the female, or the more feminine of the two males, needs to keep warm longer in order to reach its complete maturity. As is seen with the eggs that a hen sits on: all the chicks do not hatch at once, but at various intervals, in accordance with their sex or complexion, with the mother hen's closeness to the egg, and with the part on her body that is the warmest. So let people cease being amazed at how the same woman can carry one child for ten months, and another for less than nine, namely, for seven months.

It only remains to be seen how one must count the months of pregnancy, and what the count is based upon. Hippocrates teaches us to count by weeks when he says that the child is perfect, mature, and ready to come out after three dozen weeks, that is, two hundred ten days, which comes to seven months at thirty days a month. Lawyers consider the child legitimate when it is born within this period of time in a legitimate marriage, and this is based on the authority of the most learned Hippocrates, as Paul maintains in the *Digests*.[10] The same author gives four dozen weeks to those of the second rank, that is, two hundred eighty days, which comes to just over nine thirty-day months. It is the same as when he gives them seven forty-day spans, as seven times forty days comes to two hundred eighty, that is, nine months.

I do not see why these numbers, either simple or multiples of seven, have the force that some people think or make the fruit viable at seven months. Nor do I see the reason for alleging that an eight-month child will not live (especially when it has made efforts to come out and be born during the seventh month and was not able to) because it is tired and weak, or alleging that if it returns to make the same kind of effort the following month, it will die. For one could just as well say the same of the tenth and eleventh months, which are nonetheless considered to be viable. Is it not likely that the child will have made efforts to come out during the ninth month (which is a mature term) and then be born in the tenth? And likewise, the one born in the eleventh to have made an effort the preceding month? For it has been observed that with each month the child manifests some unusual movement after the sixth month. As for the tenth and eleventh months, the child has but to reach them without being born for one to say that it is a ten-month or an eleven-month baby. This is what Hippocrates says in his *Octimester*.[11] And Pliny, following him, says that a woman will sometimes carry a child up to the beginning of the tenth or of the eleventh month.[12]

In closing the discussion, I would dare say, even though it seems at odds with Hippocrates's reckoning, that the months must be understood as lunar months and not solar: that is, of twenty-seven or twenty-nine days rather than thirty. For a woman has but to enter the seventh, ninth, tenth, or eleventh month to make the child viable, which would not be the case if each of these months had to be complete, thirty-day solar ones. Furthermore, it is more reasonable to have the moon guide the count since it is the moon that guides the woman's menses, which are the rule of conception, of the child's nourishment both in and out of the womb, and of all its progress. This is also why the ancients always had recourse to the moon, which they variously called Diana and Lucina when it was a matter of childbirth. For under a certain point of her position we are all conceived, and under a similar one we are all born in the order of nature, if the birth is not hastened or delayed by an evil dominion. And thereupon are founded genethliacs, the casters of nativities, when they observe the planet that was ascending at the moment of birth.[13] But their calculations are not centered on its birth but on the planet that corresponds to the child and that was ascending during its conception, especially since it is at that precise time that an impression can be made at such and such an inclination, not once the child has been formed and quickened, and even less when it is born. Otherwise, the lapses that are said to hasten or delay childbirth would be the cause of another constellation that must be firm and fixed, or it will have no power.

CHAPTER III
THAT IT IS NOT POSSIBLE TO KNOW FROM HER URINE IF A WOMAN IS WITH CHILD; AND WHAT THE TRUE SIGNS OF PREGNANCY ARE

It is certain that one cannot know for sure if a woman is pregnant or not from her urine. For even with other conditions, both in men and in women, whether in a healthy, ill, or neutral state, this sign is more untrustworthy than any other. Now, the urine of a woman who is suspected of being pregnant cannot indicate anything other than ordinary retention of her menses, on the basis of which conception is presumed. But what good does it do for the physician to understand and know that she has not had her flowers, since the woman is much better able to know this herself, and with more certainty?

From such evidence it cannot be inferred or concluded that she is pregnant, for in many virgins these purgations are often suppressed; also, some pregnant women do not stop having them, at least for the first few months, and a few throughout the entire pregnancy. Besides, a pregnant woman can have several indispositions that make it difficult to ascertain the main indication of pregnancy (if it is present), such as a headache, cold, cough, indigestion, backache, etc. Furthermore, she must also not have had any fruit, salad, milk, meat, peas, asparagus, cabbage, artichokes, truffles, or anything else out of the ordinary in her diet that might change the color, consistency, or contents of the urine.

I leave aside the infinite diversity of this excretion, noted by the most diligent of physicians, not only according to the particular complexion of each woman and of her age, but also the season, region, dress, living habits, livelihood, interests, and countless other things, of which the slightest thing (so to speak) can alter and change the urine of the same person, not only on a daily basis, but also on an hourly one, and even from moment to moment.

Therefore, what assurance could one have of conception through urine examination? One must recognize that urine conveys rather faithfully the state of the veins and arteries of the entire body, provided it is not bloated because of rheum, which causes dripping from the head into the stomach, and provided there is no foreign matter included in it which might change its color, odor, contents, and other natural qualities, as I have amply demonstrated in my treatise on urine, composed in Latin.[1] In it I also pointed out how urine is not very trustworthy in revealing the condition of the organs above the liver, especially because most often the different organs are in different conditions, and sometimes only one of them is ailing while all the others are in good condition. For urine is drawn from all the organs of our bodies by the singular virtue of the kidneys, and the portions that come from each organ, by passing through the smaller vessels, finally reach the vena cava, which is the great canal in which all the portions of the serosity[2] (which will be called urine) are mixed and blended, and still more so, passing from the pumping vein into the straits of the kidneys, where it is strained.[3] So the signs and reports carried by the portion coming from a particular organ are clouded by the others, just as the report from the ailing organ will be erased by all the other portions from the rest of the body, which as a whole is quite healthy. Thus, not much trust (as they say) must be put into urine examinations.

Still, the most solid conclusion to be drawn pertains to the condition of the organs properly called urinary, which are those situated from the liver on down or, rather, below the pumping vessels: namely, the kidneys, the ureters, the bladder, and the canal shared by sperm and urine, and which is connected to the conduits,[4] or sperm pouches. The urine can indicate very clearly the state of all of these, especially in venereal gonorrhea, commonly called a burnt piss. And the urine shows with even more certainty the nature of these organs when there is something abnormal that it dissolves and carries off with itself, because of which it sometimes becomes cloudy and thick, slimy, or white as milk; other times pus laden, bloody, sandy, or full of hairs and small worms, little pieces of flesh, scales like bran, small bits like rough flour, little stones, and coarse sand. When these things are in the urine, they give absolute indications of the organs from the kidneys on out, through which it has passed.

I suspect that someone will think this argument is made for those who await the decision on conception through urine examination.[5] For it seems that urine comes from the womb no less than it comes from the bladder, since women piss from their shameful parts, in which copulation and conception take place. "Doesn't the urine come," one will say, "from the same place as the child? Why wouldn't the urine give a sure sign of this just as it does of the other places it goes through? We also notice that when a woman is about to deliver she lets out a lot of water, which is, properly, urine coming from the womb." I respond by saying first of all that this water comes from the womb and is urine for the most part; but it is the child's, and not the mother's. This water was retained and held in the skin of the secundine, which, breaking open when the little one comes out of it, lets this water flow out and helps to make the passageway more slippery. But the mother's urine, both during her pregnancy and when she is not pregnant, does not come from the womb or touch it in any way whatsoever. The urine is carried into the bladder by the ureters, just as it is in men, and from there it comes out through its neck into the large canal of the shameful parts, which is like a sheath of the virile member, very remote from the womb, which is much further back and much deeper.

And so laymen are mistaken when they think urine comes from the same place as the child, and that it can carry in it some sign, such as when they say: "When there is a cotton flock or a wool flock floating in the middle of the urine, she is with child!"[6] There would be an awful lot of pregnant men if such were a true indication!

"But there are some who nevertheless predict it when it is the case," someone will say. "And there are a lot of witnesses of this."[7] I say it is by coincidence (just as in heads or tails, and other games of chance) if they predict correctly only by inspecting the urine, and if they are lucky enough to happen upon a correct prediction often, it is just like being lucky in throwing dice. They could predict just as well without even seeing the urine, which only serves as an amusement with which to fool people all the better. In fact, very often these diviners are deceived by being given a man's urine said to be from a pregnant woman and, for good reason and most fittingly, are roundly mocked.

Upon what, therefore, is one to found the knowledge of a woman's pregnancy, since the urine carries no assurance? I more readily stop to pay attention to women who have had experience, who have conceived often, and who are mothers of several children, from whom one must accept what they have often experienced regarding the changes that a pregnant woman feels in her body, both in her belly and breasts. There are many other signs that are, unfortunately, neither ordinary, necessarily consecutive, nor demonstrative (in Greek we call them pathognomonic); they do not come from a particular indisposition in women but are equivocal, that is, they occur in other conditions besides pregnancy and do not occur in all pregnant women.[8] These are discharge, lack of appetite or appetite for unusual or absurd things,[9] vomiting, weakness, nausea, stomach pains, moodiness, excessive spitting, headaches, backaches, swelling of the legs, fatigue, and heaviness throughout the body. There is nothing in all of this that a virgin could not suffer from, not only separately but all together, because of the suppression of her flowers; and she will even have milk in her breasts, which is more telling than the rest, as we shall prove in the third chapter of the fifth book.

Is there, then, not a single sign of pregnancy of which one can be certain, so that a woman can be at ease or know when she is liable to hurt herself and to miscarry? Here are the principal signs that women should watch for:[10] the man's sperm is retained, which otherwise descends and flows out a short time after copulation, and in that instant the woman feels a slight tightening and contraction, with a little stiffness, like a shuddering deep in the place where her womb is, just as we sometimes feel, after having pissed, a little shivering because of the contraction of the bladder. And even up and down the spine the woman feels colder than elsewhere.[11] Soon afterward the waist becomes thinner at the navel, as if worn away. When it comes time for her flowers, instead of having them, her breasts harden and sting a little because of the blood that is dilating and making them swell. At that time she can say that her basket is full.

To be more certain about it, one gives diverse tests, which I shall not discuss for long here, because they are not certain and because of the danger in which one can put the child. For this reason they are hardly worth anything except to strumpets and lewd women who fear not offending God and killing their children in order to satisfy their lasciviousness. I would pass over these tests in silence if they were not already altogether too well known, and thus, in listing them I will not be teaching people to do evil. They know still others that are far worse, these wicked women. And I am forced to talk about this in order to warn the good women not to put themselves in danger of losing their fruit because of wanting to be sure of being pregnant through such means.

The common tests are in Hippocrates:[12] "Give a woman hydromel made with rainwater when she goes to bed. If she is pregnant she will have sharp pains," says Hippocrates—Also, let her insert down below a strong smelling and penetrating perfume while thoroughly wrapped in blankets: if the odor does not come to her nose, she has conceived; or likewise, if having put a clove of garlic into the shameful parts upon retiring, she does not have the taste of it in her mouth the next day.

CHAPTER IV
IF THERE IS CERTAIN KNOWLEDGE THAT THE CHILD IS MALE OR FEMALE, AND THAT THERE IS ONLY ONE, OR TWO, OF THEM

As for ascertaining whether the child is male or female, Hippocrates informs us in one aphorism[1] that when carrying a male the woman has better color and, in another,[2] that such a child is more toward the right side. This, it must be understood, is what happens most often. For naturally a woman is healthier and more nimble with a son than with a daughter, if there is no condition other than the pregnancy, as should always be the case.[3] For because of some other illness combined with pregnancy, the mother could be benumbed, heavy, and dejected. Otherwise, she has a clear countenance, a more vermilion color, and blithe and lively eyes, because the son, being hotter by nature, adds to the heat of the mother. But as for the right side or the left, I see no reason for it, especially since the womb is in the middle of the body, sitting upon the pelvis; and having no small compartments to the right or to the left, it is occupied completely by the child. For which reason also it is normally carried square in the middle of the belly; or if it leans a little more to one side than the other, it is only following the inclination struck by the woman to go to that side more often or more usually.

Still less certain are the signs given by laymen: that if it is a boy the woman has a better appetite, she feels the child move within three months, she has a pointed stomach, all her right limbs are quicker in their movements, her first step made straight ahead is with the right foot, if she is seated she puts her right hand on her right knee to push on if she wants to get up, the right eye is more nimble, the right breast gets larger sooner, and the movement of the child is on the right side, the opposite of a girl's. They also say that if one places on a pregnant woman's head (without her knowing it) a parsley plant with its root, if the first name she pronounces is a man's name, she is pregnant with a son; otherwise, with a girl. Also that if a pregnant woman lets fall into some water a drop of her milk and it sinks to the bottom, it is a girl; if not, a son. The same is said of her blood. They also give the argument that if a woman has a nosebleed, she is pregnant with a girl, due to the fact (perhaps) that her blood is more subtle and serous, or that a girl does not use up as much as a boy.

But I pay more attention to the color and consistency of the milk, which is usually more subtle and tawny when it is a girl, more thick and white when it is a boy. This is also why when this milk is thrown against a mirror or some other smooth surface it will bead like pearls or balls of quicksilver, even if this is done in full sun. Likewise, if some is thrown in water, it sinks perpendicularly because of its fatness and heaviness, which will not be the case with milk from a woman pregnant with a girl, because it is lighter and more subtle, just as it is hotter and more fiery, as we shall demonstrate more fully in the fifth chapter of the fifth book, against the opinion of laymen. Yet this milk is also more tawny and serous, just like venom (which is acrid and biting), compared to the more desirable kind. But, as I have shown above, it takes very little to alter these signs: the slightest thing can upset everything and render the most certain indications completely false.

There remains the question of whether one can know for certain that the woman is carrying two at once. It is not a matter of the womb's being divided as if into two small compartments, a left and a right; but rather, in the same space there will be two, three, or four, and up to nine, as we have shown to be possible in the first chapter of this book. From two children a mother can feel diverse movements at the same time, and the two sides will be more swollen and raised than the middle of the belly, where most often a small depression is seen. Yet, one is often misled by this, for we have seen it happen that the womb, overwhelmed with a large and restless child, will slide to one side and, pressing little by little on the bowels, will push them to the opposite side. There then seems to be another child that is not moving at all, and one will say that it is a girl, the other a boy; but more often there is nothing but a big fat daughter, posing as the two, who has thus gained room on one side. One can also be mistaken because of a lump of tissue we call a mole,[4] and the Lombards a harpy,[5] which we shall treat in detail in the next book.[6] It looks like a child, sometimes off to one side. And so, there is scarcely any certitude regarding the number of children, and still less regarding the ascertainment of their sex. I shall always believe far more in this matter the children who have just been born than the greatest philosphers and physicians in the world.[7]

CHAPTER V
THAT IT IS A GRAVE ERROR TO TAKE LIGHTLY THE ILLNESSES THAT COME ABOUT IN RELATION TO PREGNANCY

There are some women who have a very healthy pregnancy, that is, who do not feel any differently than normal, so that, were it not for their growing belly, they could easily hide being with child. It is the only thing revealing their condition. Besides this, they know their purgations have stopped; then the child's movements, around the third or fourth month at the latest, make them certain of it. Such women are healthy indeed, and their fruit is strong, consuming superfluous blood in the mother's body; this type of blood is perfectly fitting.

It follows that there are no vitiated or ineffective humors, either in the child or in the mother, backing up in the stomach and other parts of the body, from which several illnesses and complications (especially in the first months) arise in people other than pregnant women who are full of foul humors. For such bad fluids in the body, unhealthy both in the body of the mother and of the child when natural purgations are suppressed, stand about and stagnate inside the belly, and from them result vomiting, nausea, loss of appetite, or appetite for unusual things according to the dominant humor, aversion and hatred for all that she used to like the most, weakness in the heart, shortness of breath and suffocation, perspiration, watery mouth, exhaustion, heaviness, and swelling of the legs. All of these conditions and side effects affect virgins failing to have their flowers at the proper time, as often as pregnant women, and, among other illnesses, causes them to hunger for unusual, absurd, inept, and bizarre things, which is called a craving.[1] Such as wanting to eat paper, plaster, ashes, coals, wheat, flour, pure vinegar, pepper and other spices, green and sour fruit, etc., and hating all other good food.

This comes about (so they say) both in virgins and in pregnant women because of foul humors retained through the suppression of their menses, which makes them desire things similar in nature, namely, foul things. And so one must not conclude from these cravings that a woman is pregnant. One can correctly say that she has cravings like those of a pregnant woman.

Now, in young girls, widows, and other women known not to be pregnant, we labor and try to heal all these illnesses because they are very painful and destroy the body. Pregnant women, however, are left to endure all of these, and the poor things must have patience until they deliver, "until the hot water takes care of everything," as women often say (that is, the bath given them at their lying-in); unless, rather, these illnesses take care of themselves, as is most often the case, when the child gets bigger and consumes all that is superfluous, good and bad.

This opinion seems to be somewhat correct, inasmuch as we do treat young girls, widows, and other women who are not pregnant by soliciting and promoting their menses. For in arresting the cause, we arrest the effects. Take away what is evil (the obstruction of the uterine veins), and all the accidents will cease, which one will combat and try to heal in vain as long as their cause is unassailed. But in pregnant women we cannot (at least we should not) use such remedies, since provoking menses would be tantamount to performing an abortion, a scandalous, inhuman, and damnable act. For it is a veritable homicide and most cruel killing of an innocent little one. This is why it seems that these poor women must, in all necessity, endure all these ills, and why it is not lawful for the physician to intervene in any way.

Yet we see that all the most learned and famous in our art, Aetius, Paul of Aegina, Rhazes, Avicenna, and their followers, have not neglected such ills but have taught us to treat them in pregnant women.[2] Have they done an evil thing? Or do we do evil in not imitating them? The ignorant layman keeps our hands tied and hinders us from helping them.[3] It would indeed be most evil (and here is where laymen found their opinion) to provoke menses in a pregnant woman, since their retention is necessary for conception and pregnancy. Nor must pregnant women be bled if there is no necessity other than the ills we mentioned above, such as a continuous fever, pleurisy, quinsy, or some similarly acute disease, fatal for the most part in pregnant women.[4]

Purging is similarly suspect, as are the strong remedies used by Galen and Hippocrates,[5] who did not know of the benign and gentle ones since discovered. Now, the minor discomforts of pregnancy do not need these strong preparations and remedies used on serious illnesses that confine one to bed. But light and gentle remedies, purgatives, and others as well are in no way forbidden in these instances but are highly required and necessary, in my view, following the opinion of the most learned and expert ever to have written on medicine.

What good does it do to make a pregnant woman suffer from vomiting, which racks the belly and rib cage and puts the child in obvious danger of a miscarriage? A light medication, such as rhubarb, which is very relaxing for the heart, will relieve her of this strain, without unsettling or upsetting anything, by ridding her of choler and other foul humors that upset the stomach and stop it from keeping down food, thus causing the mother and the child to be poorly nourished. What good does it do for the mother to suffer nausea, queasiness, and disgust for all sorts of good food because of foul humors encumbering her stomach when they can be expelled once and for all? Is it not a mean cruelty to make her suffer these and similar ills so long when they can be relieved so simply, without harming the child? What am I saying, "without harming"? It would do it, as well as the mother, inestimable good.

For look at the results of allowing these excretions to settle and stagnate—the cause of all the ills from which a pregnant woman suffers. First of all the mother will fast, of necessity, for she can eat nothing worthwhile, or if she eats, she vomits it immediately. The child takes the best nourishment it can, as long as it is able to choose and draw good blood from among the foul and the filthy. But when there is no more good blood to be had or very little, it is forced to feed upon what it can find, for necessity drives it to take its fill, be it hay or straw (as is said proverbially),[6] as does the body of the mother; thus, both suffer from it.

Would it not be better to get rid of these impurities so that the woman, regaining her appetite and no longer vomiting, will have enough good food, both for her own body and for that of her child? It must not be feared (as I have said) that harm will come to the child from a mild remedy, such as rhubarb, which, because of astringency, will actually strengthen rather than weaken it. What is there to fear from these remedies when there are pregnant women who do not miscarry after the greatest of strains, such as falls, blows, fits of anger, and so forth? There are enough of them who are not afraid to ride a spirited horse, dance a round, a pace, or a leap, while pregnant up to the throat. And would they be afraid of a remedy that does not upset them at all or only very slightly, and that has the advantage of making vomiting stop and weakness in the heart as well, along with heaviness, fatigue, shortness of breath, and other unpleasant side effects of pregnancy, when the body is full of foul humors? If a woman is prone to excitement over small matters, she should refuse to be suspicious of such remedies. For I shall continue to affirm that the strain of vomiting and lack of nourishment will sooner cause her to lose her child than will these light purgings.

And so the conclusions are most evident, as I have shown. For we only fear purgings, along with Hippocrates and Galen, because of the agitation and the upset caused by hellebore and other strong drugs, as is said today of antimony. But vomiting in a pregnant woman shakes the body imcomparably more than our light remedies. And as for bloodletting, we, along with the aforementioned authors, do not fear it except for the harm that a lack of blood could do to the child thus deprived of its nourishment, for which reason it is forced to evacuate because of a lack of supplies. And does one not deprive it of its food when the mother eats nothing, or much less than the child requires? It seems to me that one does great harm to pregnant women by letting them suffer thus, enduring pain when it is simply not necessary. There is, in addition, another disadvantage in not acting: the child will never be as healthy as it would have been because it fed so long upon nastiness and filth. For its body will be more inclined and likely to accumulate similar uncleanness, and it will have to take a hundred drugs in its lifetime because of one that was kept from it when it was in the belly of its mother.

CHAPTER VI
WHY IT IS SAID THAT WHOEVER REFUSES A PREGNANT WOMAN SOMETHING WILL GET A STY IN HIS EYE

An *orgeol* [barleycorn],[1] in the language of this region,[2] is a small tumor or swelling, longer than it is wide, the size of a grain of barley (whence its name), growing on the corner or edge of the eyelid. French, corrupting the word, gives *orgueil* for *orgeol*,[3] which corresponds to the Latin *hordeolum*, and the Greek *krithe*. It is not a serious ailment and is more bothersome than painful. It goes away by itself, most often disappearing without a trace; sometimes it becomes infected and will spurt out some foul matter. When you notice one on somebody, you usually say, "You refused a pregnant woman something!" Or if a person refuses a pregnant woman something, you say, "You're going to get a sty!"

These are little jests, quips, and familiar sayings meant to encourage earnest people to comply in what they can and ought to give to pregnant women, who are in danger of miscarrying because of an immense desire for something they cannot have. One thus threatens children who play with fire to dissuade them from doing so (because they risk burning something up or starting a fire somewhere in the house) by telling them that it will make them wet the bed, which they fear tremendously, knowing that they would be whipped if they were to piss in bed. Similarly, people tell children that touching the flower of the red poppy (called a *lagagne* in Languedoc because it causes people who look at it very closely, if they have tender and delicate eyes like a child, to get red eyes and become blear-eyed) will make them wet the bed. To those who are still smaller, one often says that if they drink while eating their soup they will not be able to see anything when they are dead. The reason is to dissuade and hinder them from interrupting the heat of the soup, which does their stomach good. Also, because going suddenly from hot to cold damages their teeth and gums, which are very sensitive in children. Thus it is with the sty in the eye (or on one of the eyelids), which the gullible are afraid of getting if they refuse to give a pregnant woman what she has a huge appetite for, as if the sty were a punishment for the danger of miscarriage to which they exposed the woman.

For, in truth, a miscarriage can come about (in a woman susceptible to it) from a strong desire, or from spite and anger over not being able to obtain what she inordinately desires, no less than from a great fit of anger, joy, sadness, or some other emotion. For such perturbations sometimes cause sudden death in little women and old men, whose link between soul and body is very weak and easily broken, as we have shown in the first book of laughter.[4]

How much more easily could emotions be the cause of death in a child, or of miscarriage. Emotions and perturbations of the soul are like the winds and storms that agitate the water of the sea and make it move to and fro with such violence. Our emotions can thus so greatly move and upset our humors that they will burst forth on all sides. And so because of a fit of anger or spite, the menstrual blood that had been retained for the child, now disturbed and driven out, surrounds and carries off the child just as a torrent sweeps away a huge boulder.

This is why it is very dangerous to refuse a pregnant woman something, especially if she is crazed and among those who have terrible tempers and difficult pregnancies, or who on the contrary are too patient and restrained, smothering their appetites, in which case the hunger and strong desire swell even more for having been kept hidden. Marcus Aurelius tells of Macrina (the respectable wife of Torquatus, the Roman consul), who, being with child, died suddenly from an excessive desire she had to see a one-eyed Egyptian (his eye was in the middle of his forehead), who was walking in the street in front of her house and whom she did not dare look at because of the Roman custom of not being seen at the window (nor of even going out of the house) in the absence of her husband, who was off fighting a war against the Volsci.[5] The senators greatly regretted the death of such a virtuous woman. Sometime later, remembering this misfortune, they granted to Roman ladies who had proven themselves generous to the republic in times of great need the privilege of not being refused anything of a decent and legitimate nature they asked for when pregnant.

The particular generosity moving the senate to grant them such a privilege was of the following nature: Camillus, a most renowned captain, leaving Rome to go to war, made a solemn vow to Berecynthia that he would erect in her honor a silver statue if he returned victorious. Having obtained his wish, there was not enough wealth in Rome to pay for the statue. When they heard this, all the ladies spontaneously went up to the capitol and generously put at the senators' feet all their rings, jewels, chains, necklaces, bracelets, belts, wedding bands, brooches, and trinkets, along with all their stones. One of them named Lucina, on behalf of all the women, begged the senate not to esteem the treasure they gave so freely (to make a statue for Berecynthia) as much as their husbands and sons who had risked their lives to win this victory. The senate, moved by this magnificent graciousness and generosity, rewarded them with five great privileges, among which was the aforementioned one: that one could never refuse a woman with child anything legitimate she might request.[6] Second, that from then on women would be honored at their funerals with processions, orations, and epitaphs. Third, that they would have the right to sit in the temples. Fourth, that every woman could have two costly dresses without the senate's permission to wear them. Fifth, that they could drink wine in the case of extreme need and grave illness.

This is why, ever since then, the wishes of pregnant women have been respected, and why this little saying was invented, that whoever refuses a pregnant woman anything will get a sty in his eye, that is, some visible punishment (such as some sign on the face), if but a small one.

CHAPTER VII
WHY WOMEN WITH CHILD ARE ADVISED TO PUT THEIR HANDS ON THEIR BOTTOMS WHEN THEY CANNOT SATISFY THEIR CRAVINGS IMMEDIATELY

A thousand stories are told of the visible marks on children's bodies, all attributed to a strong desire or appetite not fulfilled or satisfied by the mother when she was carrying the child. Some have a mark like a cherry, others like a strawberry or blackberry on one of their lips or on their noses or some other part of their bodies. Still others might have a mark that looks like a currant, fig, melon, cucumber, or some other fruit, on the thigh, leg, foot, or elsewhere, and it is due to the mother's having a craving for such fruits when they were not in season, which is why she could not have them.

Another child might have a mark like a hare's lip or a snout recalling the head of a herring or a lamprey because the mother wanted one or the other of these fish and could not have it. A little girl was seen with a perfectly formed ace of spades in the middle of her cheek, black in color. Another child had a perfectly shaped sage leaf. On one was seen something like a sleeping rabbit across his neck, while another had in the same spot the marks of the holes made by a lamprey. On still another one could see on the middle of the cheek a red rose so perfectly portrayed that the leaves could be distinguished one from the other. One child might have in the middle of each arm a thing having the shape of a sheep's footprint; another, right next to his fundament, a mark like the horns of a stag.[1]

They tell of a woman from Auvergne who had a great craving for the skin of a butcher, who left his white fleshy arms uncovered. The woman, driven by this mad hankering, told the butcher, who was so striken with sympathy that he immediately cut a piece of muscle from his thigh and gave it to her.[2] The woman was overjoyed, ate it raw right on the spot, and was thus quite satisfied. She had two male children, one of whom had a piece of flesh on the end of his lips, and the other always had his mouth hanging wide open. The latter (as they explain it) did not get his share of the meat, which hangs from the mouth of the former, so he keeps his mouth open as a sign of the desire for it that remained in him. They say he is completely witless.

I was told of another who has a red spot on his hand that becomes redder and darkens in color noticeably during the grape harvest. They say that when his mother was carrying him she had a great desire and extreme craving for new wine on Saint John's day when it was impossible to have some. Another one has several white spots on his neck and all down his chest, like drops of milk (which could be white morphew),[3] because (so they say) his mother, when carrying him, suffered a strong desire for milk at a time when it could not be procured quickly enough.

Now, I do not wish to refute completely the truth of these reports, which are most often poorly reasoned stories and are as ill founded as the one about the woman who told her husband she got pregnant during his absence by eating snow when she had a violent craving for currants. For just as in a child that is already bigger, and in a grown man, there appear diverse tumors and lumps[4] of different kinds, so, too (and even more easily), can these marks be made right from the first formation, no more or less than six fingers or six toes, or a toe split in two, or two fused together halfway to the ends, as in all the children of Monsieur de Joyeuse, the king's lieutenant general in the region of Languedoc, true father of the region and people.[5] And these marks or spots that are not tumorous are the same as the ones we discussed in the third chapter of the second book, namely, morphews, not only red, but of every color, white, black, sallow, green, etc. If they resemble the shape of something, they are named after it: cherry, melon, herring head, lamprey neck, ace of spades, sage leaf, sheep's foot, milk spot, etc.

I nevertheless will grant that a vivid imagination and apprehension in the mother can have a great influence on the body of the child in imprinting upon it some mark; but this is mainly at the moment of conception, or during the time the child is being formed, which can be as long as a month according to Hippocrates: thirty suns (that is, natural days) for formation, sixty for movement, two hundred and ten for completion.[6] This is when the woman has her greatest cravings, as she has the greatest amount of excretions still retained.[7] This is why some affirm that if at this time a woman with child wears a nosegay in her bosom or behind her ear, or some other place touching her skin, it will appear on the child, as if she wanted her child to have a nosegay where she was wearing it on herself.[8]

In this first month, devoted to the formation of the child, the imaginative faculty has a fair amount of influence, of which I have given several examples and causes in my preface to the second book on laughter.[9] But when the child is already completely formed and is moving about, it is quite strong and no longer subject to these influences (in my opinion), if it is only a matter of the mere imagination of the mother, however vivid and strong it might be. I say "mere imagination" because if there is some illness in the mother it could well appear on the body of the child, in the same spot, as seen recently in the city of Nîmes: a woman had a carbuncle on the right shoulder that made her miscarry in the eighth month a daughter who also had a carbuncle in the same place.

Let us now come to the question of a woman putting her hand on her behind if she cannot be given immediately what she desires. Ignorant people are of the opinion that if during this fancy or craving she touches herself on the face, the nose, the eye, the mouth, the neck, the breast, or some other part of the body, there will appear on the corresponding part on the child a mark of what the mother had a craving for. And so in order that this spot be hidden, it is better that it be printed on the buttocks or some other part that clothing will cover.

Now, if the foregoing, which is feared, is true, then this is good advice. But it is folly to think that, if there must be an imprint on the body of the child, it would be in the same place that the mother touched first. For in all this there is no reason whatsoever, nor evidence; or else it would be necessary for the mark to appear first on the mother on the part she touched, and from there it could be communicated, as we said above about a carbuncle. And I think there are no more occurrences or evidence of such things than there are reasons for them; it is, rather, a notion held by ignorant people without any basis other than (as they say) regional hearsay.

CHAPTER VIII
CONCERNING WOMEN WHO EAT GREAT AMOUNTS OF QUINCE MARMALADE DURING PREGNANCY SO THAT THEIR CHILDREN WILL BE QUICK-WITTED, AND SUN-DRIED RAISINS SO THAT THEY WILL HAVE SHARP EYESIGHT

It is commonly known that quince marmalade tightens and restricts loose bowels, strengthening the retentive virtues of the stomach and intestines with its astringent qualities, which are most apparent. Because of this, women have since come to the opinion (I believe) that quince marmalade can also help the retentive power of the brain, which we call memory. And for this reason they say that quince marmalade makes the child quick-witted, especially while in the womb, for being soft, it readily accepts all impressions.

One calls being "quick-witted" the ability to understand quickly and to retain well what is seen. In order to understand, softness is needed rather than astriction, which is rough and dry. But understanding something counts for little if it is not retained for a fair amount of time. Now, a child is so soft that its impressions are about like writing in air or in water, or (to give a better comparison) like what is imprinted in dough or very soft wax. It is but a waste of time; some firmness is needed in that which is to last. And so, the child will have as though no retention at all, until its body dries a little. This is why people say that quince marmalade (an astringent and desiccant) makes it quick-witted.

But is this a good thing? By no means, and for many reasons. First, the mother, who is usually constipated as a result of her condition, becomes unfortunately even more constipated. Second, quince marmalade, in its influence on the child, does nothing that can be considered useful, or that another desiccative food could not do just as well. But it is not good for the child to become dry. Natural softness helps the growth of its body, which stays small when the dough is very dry.[1] Besides, he who is born drier gets old and to the end of his road much sooner—something everyone wants to avoid as long as possible.

We also note that these children who have so much intelligence do not live long. This is why common people commonly say, "He was not meant to live, for he was too smart." The reason is that the principal activities of the active and very agile mind dry out the body, which is almost ceaselessly taxed by it. And when the body is made dry, the mind does become sharpened, but not for long. Nature, therefore, must never be forced in any way, and since it is a child's nature to be soft and moist (because this makes it grow more quickly and live much longer), his quick wit is not to be a concern. It will take care of itself well enough if the body avoids excesses. For the principal action of the temperate man is prudence, as Galen says in the first book of *Complexions*, or *Tempers*.[2] And he is most temperate if he is wellborn and well-bred, having been engendered and conceived from very healthy parents.

Excellent memory and a quick comprehension are not as praiseworthy as many people think. Both abilities are intemperatenesses of the brain, the one too dry, the other too soft. And so such brains are not those of the most wise, as we have observed in several people with monstrous memories (if I dare so speak), yet who were at the same time imprudent, dazed, harebrained, and witless as the first bell of matins.[3] Of such people it can well be said that they have great minds, that is, for understanding and retaining everything they wish—nothing gets by them. But in thinking, reasoning, and judgment, they are less able than several others with a poor or faulty memory. The temperate man (who is consequently also prudent) has all his faculties moderately developed, none to excess, understanding quickly enough, remembering rather well, and perfectly wise. One must not, therefore, be excessively concerned about being quick-witted, or having a great memory, lest judgment (the most important activity of all) be somewhat hindered by them.[4]

Turning to the other question, that of sun-dried raisins, or what we call *paserilles* in Languedoc (from *uve passe*[5] in Latin, and the most renowned are those of Damascus in Syria), it is likely enough that if a pregnant woman has them often her child will have sharper eyesight. This is not because of any oxydercical property (that is, sharpening the sight) that might inhere in these dried raisins, but because they are extremely nourishing and make for rich, pure, and clear blood; when the child is nourished with such, it will have quick and unfettered perception because of the clear, nimble humors furnished, much more so than if the child had been nourished on a coarse and heavy blood.

That raisins are a good and healthy food, I have amply shown in the *Matinées de l'Isle d'Adam*,[6] and the experience of those who eat them regularly bears witness to it sufficiently. I have certainly seen several skinny, gaunt, benumbed people who, by taking this food, in a short time regained marvelous health. So one does well to exhort pregnant women to eat them often, especially those who are rather squeamish, for one will rapidly eat enough of them sooner than meat and soup.

Very close to this question is the saying that the first morsel of food goes to the child, which we shall treat in the next chapter.

CHAPTER IX
WHETHER IT IS TRUE THAT THE FIRST MORSEL OF FOOD THAT THE PREGNANT WOMAN EATS GOES TO HER CHILD

Ignorance of anatomy makes people come up with many absurd and ridiculous sayings—impossible things.[1] For example, I heard a nun, bragging about the beauty of her skin when she was healthy and younger, say that if she drank red wine it could be seen going down through the veins of her neck, so white and clear was her complexion and so delicate her skin. She did not know that wine does not go down to the stomach through the veins, but through a tube called the esophagus, which is attached to the end of the throat, and that it is impossible to tell the color of the wine even if it were to pass through the veins, since one cannot see the color of the red blood that they do contain. I have heard soldiers say that they have seen an eye come out of a wounded man's head, rest in his hand a moment, and go back into its socket so perfectly that he saw out of it just as before. And similar stories about a nose cut off completely and fallen to the ground.

There are some who tell still other stories of an impossible nature and that serve only to make people laugh. Such might be said of the one being proposed: that the first morsel eaten by a pregnant woman goes to her child. For many people, not knowing anatomy, think that the child in the womb eats and drinks like the mother, and they do not know that it is nourished solely by the blood it draws through the navel. For it lives in the womb like a fruit on a tree, drawing the nourishing sap from its mother plant through the stalk, or stem. The child takes nothing through its mouth until outside the womb, and the first thing it takes in then is air, which until then it had not inhaled.

Even if the child in the womb were to use the same food as the mother, as many believe, it would still not necessarily follow that the first morsel would go to the child rather than the last, or any other part of the food. For everything that is eaten and drunk by the mother mixes together in her stomach, is reduced and digested, and then stays there (if the stomach is good) until it turns into a substance that is completely uniform in color and in consistency, called chyme. It is like very thin barley pottage without lumps. Then, when the stomach has had its nourishment and its fill, it sends the remainder to the intestines, whence the liver draws off through the mesentery veins what is most apt for conversion into blood. With this blood the child is finally nourished.

It is true that the liver can, if necessary, take from the stomach a certain amount of what it has not yet received before everything is digested and reduced; this is done through the veins common to these organs and the stomach, through which the hungry stomach draws from all quarters any humors that it can procure. But there is not the slightest plausibility or possibility that the first morsel goes directly to the child. For it is nourished solely with blood, as they say, and there is always blood to be had in the body of the mother, especially around the womb, where it tends to be more copious.

It is also true that the child makes the mother hungry when it begins to get bigger and to consume a lot of blood, which is why the mother is forced to eat more than usual; otherwise, she gets weak and faints easily. This is not to say that the child draws the food to itself, or that when there is no food it turns for nourishment to the blood, which then will begin lacking in the mother, and that for this reason it is necessary that the mother be better nourished. Rather, it is necessary for her to be better nourished so that she will have more blood, which will suffice both for her and her child, which is nourished with blood just as if it were one of the mother's limbs.

Why, then, do people say so foolishly that the first morsel goes to the child? Is there no reasonable basis for this saying? We believe that most of the sayings and expressions of the populace come down from the philosophers and other divine personages who have taught people to survive. Did this saying not come from such a one? Or is it from simple ignorance of anatomy, as we said in the beginning? People certainly betray ignorance in this saying, but it could also be the case that it was given to them in this gross form, given their limited capacity, to exhort pregnant women to take proper nourishment, as is very necessary, so the child will not lack good blood and will thus be robust and healthy without harming its mother.

Why do they speak about the first rather than the last or the other morsels? It is easy to understand that one does not mean only the first morsel or mouthful of anything whatsoever, but of the first food, such that if only mutton and beef are to be served, the pregnant woman must start with the mutton. And if there is also a capon or a partridge, let her eat these rather than the mutton, and so on and so forth with other foods that are still easier to digest. Let her begin with a good soup and leave aside fruit, salad, and other Spanish foods.[2] For if she heeds her appetite's fancies and begins with chitterlings, sausage, blood pudding, salted anchovy, or sardines, it is to be feared that she will fill up on these foods that are more toothsome than wholesome and afterward will not be able to eat something better. This is why one does well to advise her to begin at least with some good food, and in order to convince her of this, one says that the first morsel goes to the child. For it is known that mothers are naturally more careful and attentive to their unborn child than to themselves, for which reason one cannot find a better way to get them to nourish themselves than by telling them that it is good and necessary for the child.[3]

END OF THE THIRD BOOK

THE FOURTH BOOK OF POPULAR ERRORS CONCERNING CHILDBIRTH AND LYING-IN

CHAPTER I
THAT THE PUBIS DOES NOT SEPARATE TO FURNISH PASSAGE TO THE CHILD

As I said in the last chapter of the preceding book, ignorance of anatomy is the reason for many absurd and ridiculous sayings, such as maintaining that the pubis (in Latin *os pubis*) opens and separates to allow the child passage.[1] Laymen cannot understand how such a large body is able to come through the ordinary passageway without considerable violence. Indeed, this tightness is the reason for the terrible pain the woman feels in delivery, especially with the first few children. Once it has been opened several times it will not hurt as much. This is why some people say that women who marry later in life, or who are for other reasons older when they have children, suffer most: their bodies, being harder and drier, do not allow the bones of the pelvis to separate easily, which is why their children often die in the passageway. Others say that the matrons and midwives of Genoa, in order to avoid these difficulties, break these bones on their daughters right at birth so they will not have any pain when the time comes for them to have children.

This is nothing but foolishness and lies, proceeding from the grossest ignorance ever known. For one must understand that the pubis is the conjunction of two large bones forming the flanks on each side, to which the thighs are attached. This conjunction is made by means of a tendon or cartilage holding them so firmly together that it is impossible to separate them without severing it. This can be understood immediately if they are seen in the open, as when we perform a dissection. And as for breaking them (as one would a capon's, or some other fowl's, in order to make it seem fuller and a better bargain), this cannot be done without considerable damage to the organs underneath, namely, the bladder, the womb, and the rectum. Furthermore, crushing these bones would cause more complications both in pregnancy and in the delivery than any comfort gained because of the pressure applied internally, unless they were to spring back out later and remain separated. But I do not believe that this can happen. Besides, there is no need for them to open, as we shall soon argue.

Now, where did this story about the Genoese women come from? There is not a single popular and common error that does not have some foundation to it, which is the source of its inaccuracy. It is (in my opinion) because the women of Genoa ordinarily have easier deliveries than others, just as they say. This is why people think these women have a more open passageway, and from there they fabricate the means mentioned above. I would more readily maintain (knowing the honor of those who are chaste and decent, for there are everywhere some of both kinds) that Genoese women, *donne senza vergogna*,[2] as the proverb goes, on the whole lascivious and prodigal with their honor, render themselves, through the frequency of love's games, more apt and quick in their deliveries. For whores are kneaded, as it were, by many insatiable lusty men, and their shameful parts are so much in use that it is easy for the child to come through the well-worn passageway. Also, they work their crupper so much, the principal part in this matter—I mean in childbirth, as will be understood shortly—that when it comes to having a child, the crupper is very supple and gives readily. Other women, who move it about much less often, have a stiffer one, and especially old ladies, who carry children less often than younger ones, even when married, which is why they hold off for a longer time. If they have more pain with the last children than with the first, this is the reason.

Likewise, women who are married off when they are older have a lot of pain in childbirth. This is why girls are married off sooner than boys, although there are several other reasons, more social than physical. Village women, and other women doing manual labor and a lot of physical exercise, and who stand more often than they sit, have easier deliveries than merchants and townswomen, who rest and sit more often and tend not to work on much besides needlepoint and sewing. This is why Lycurgus very wisely ordered the Lacedaemonian (or Spartan) women and maidens to practice wrestling to make themselves stronger in bearing all sorts of pain, even the travail of childbirth, so they might have an easier delivery.[3]

That the crupper is the principal part here, women who have given birth can attest. The principal pain they suffer (besides that in the loins) is in this part of the body, and not in the pubis, which would suffer pain mainly because of its tender ligaments if it were violently spread open, as laymen think. But it is only the crupper that suffers from violent pressure when it is pushed back in order to let the child pass through, between it and the pubis, which does not move in the least.

The crupper is a small tail made up of four tiny bones, and it is somewhat longer in certain places than in others. The Greeks named it coccyx because it resembles the beak of a cuckoo.[4] I do not know if that is why the French call a cuckold a person who allows his wife to move this part of her body to satisfy the appetites of a third party. For to call a person a cuckold because of behavior like the cuckoo would be a great error, especially since a cuckoo will not allow any other bird to nestle or lay eggs in its nest, but on the contrary goes laying eggs in other birds' nests. The term comes from the hedge sparrow, properly speaking (some call it a *curruca* in Latin),[5] which is a small bird. When it has laid five or six eggs, the cuckoo comes and eats them, and then in the same nest it lays an egg, which is much larger than those it had eaten. And so the hedge sparrow could well take notice (given the sizable difference) if it were the least bit aware. But it is so unaware that it takes for its own what it finds in its nest. And so it sits on it and then nourishes the little one that is not its own. They say that this is what happens most of the time—not always, for otherwise, the species of hedge sparrows would soon disappear.

From these arguments it can be seen that the husband is improperly called a cuckold in this sense. For it is the lusty adulterer who should be so named. But through the cuckoo, that is, the crupper, he is much defamed, especially when he is partly at fault. The Italians call him a *becco* for the same reason, because of this beak, which applies more suitably, than because of a *bouc* [he-goat],[6] for the word *becco* means both one and the other.

CHAPTER II
WHETHER IT IS GOOD TO MAKE THE WOMAN SIT ON THE BOTTOM OF A CALDRON OR TO PUT HER HUSBAND'S CAP ON HER BELLY SO AS TO HAVE A BETTER DELIVERY; AND WHICH ARE THE BEST POSTURES FOR DELIVERY

This argument will serve as confirmation of the preceding discussion. It is that the village women in the region of Montpellier have found that if a woman suffering the pangs of childbirth sits on the bottom of a caldron that has just been taken from the fire, she will bring forth her child more easily. We know that such a caldron, in which water was recently boiling, has a tepid bottom (cold in comparison to caldrons used for other purposes, which are fiercely hot).

This heat softens the crupper and makes it easier for it to relax, like softening fomentations also used for this purpose. But they are commonly applied with ill effects on the pubis and over the area of the womb on the front. They must be restricted to the crupper; otherwise, they serve no purpose and, what is worse, are harmful. I say that they serve no purpose on the *os pubis*, for softening it will not cause it to give in the slightest.[1] Moreover, they are harmful to the womb because they weaken its expulsive virtue, which needs nothing other than astriction. The more one relaxes the womb, the more one saps its strength for pushing out the child. This is why village women could do worse than to sit a woman who is about to deliver on the bottom of a hot caldron.

There are fewer reasons for the practice these same village women have of putting the husband's cap on the woman's belly, unless perhaps that once it is there, the belly tightens in the place under the cap, and in this case would act as a compress to aid in the expulsion. But I think it is done as a game. At least that is how it started, and since that time it has been taken seriously. And the game can be interpreted in the following way: that the husbands naturally withdraw and do not allow themselves to be present at such events; sometimes people try to force them to be there to help, and if their presence cannot be had, their cap is kept, which is put on the woman's belly as a token signifying that this swelling was caused by the man, as if he had a poisonous thorn, and that his cap applied on the belly acts as an antidote, making the swelling go down.

It is now necessary to consider various postures during the act of childbirth.[2] Some wish to be standing, held up; others, seated on a sawed-out chair that is also open in the front; still others, lying down. I let those who have tried everything choose the manner they find the easiest. I only warn that they be certain that the crupper is free and unimpeded, so that it is able to flex down freely. This would be aided tremendously if the standing method were used at just the right time, when the child starts coming out, without tiring the poor woman or causing her useless pain. For besides the fact that the crupper (as it is called) has great freedom of movement in such a posture, the child, because of its own weight, descends better and aids its delivery. There are ladies and gentlewomen who use beds, called labor beds because they are only used during the delivery of a child. They are not actually beds for sleeping but chairs open in the front, having arms and legs made for strapping the arms, thighs, and legs of the woman with belts, which are soft and wide but so strong and firmly attached (yet without hurting them in any way) that the woman is unable to move anything except the crupper. This is good and helpful, provided it is used wisely.

It is a thing of great importance to have the woman deliver comfortably, considering the danger that she and the child suffer when there is a complication. And so it is right that the matrons or midwives are called *sages-femmes*,[3] for they must be prudent and circumspect, especially when there are two or three children to come out, for they are sometimes overwhelmed with only one.[4] What would happen if there were nine, as I wrote about in the first chapter of the third book in the case of Madame de Beauville, and in Arles and Padua? I understand that in the house of Estourneau in Périgord a similar thing happened over three hundred years ago. The woman had nine male children in one burden and wanted to expose eight of them; they were, fortunately, saved (by the grace of God) because of a chance meeting with their father.[5] All nine lived and were provided with great estates, four of them in the church and five of them in the secular realm. Of the ecclesiastics, one was bishop of Périgueux and abbot of Brantôme; the second, bishop of Pamiers; the third, abbot of Grand-Selve; and the fourth, of La Chaise-Dieu. Of those in the secular realm, one was the king's lieutenant in La Réole, against the English; the second had a governorship in Burgundy; the other three enjoyed the close confidence of the king. One can still see today the entire mystery painted in a room of the castle of Estourneau, as several lords, worthy of trust, have informed me. Moreover, I know personally a descendant of this most illustrious and ancient house, called the Seigneur d'Estourneau,[6] steward of the king of Navarre, my liege, Henry the third of this name. May God grant him long and prosperous life.

CHAPTER III
THAT MATRONS ARE GRAVELY MISTAKEN IN NOT CALLING PHYSICIANS TO DELIVERIES, AND OTHER ERRORS PROPER TO WOMEN; AND THAT EVEN MIDWIVES OUGHT TO BE INSTRUCTED BY PHYSICIANS

The arrogance and presumption of some women is such that they think they understand more about conditions proper to women (such as suffocation of the womb, abortion, and childbirth) than the most competent physicians in the world. Whence, if it is a case of an illness of the womb, an abortion, or a birth, and some fever or other complication arises, they will not deign to call them in until they have tried everything they know.

I find it very good and reasonable that they share among themselves their usual little remedies, and that midwives apply the experience and skills they have acquired through practice. But if they think that physicians do not know all these things better still, they are greatly mistaken. Nevertheless, we leave to them this branch of surgery involving childbirth because it is more decent that this treatment be administered woman to woman in their shameful parts, just as we have left all the rest to the professors of surgery, to our own great relief, so that the ill might be better cared for in having now two ministers where there was only one.[1]

But the physician is not excused from knowing every single thing that midwives do, any more than from knowing every other surgical procedure. Moreover, it is most fitting that he be present everywhere, if at all possible, in case some complication arises. For all illnesses are within his knowledge and under his jurisdiction. All those who meddle in treating any illness are underlings with respect to physicians, such as surgeons, who have middling jurisdiction, and midwives, who have the lowest.

Now, childbirth is a condition from which many, both women and children, die. And from abortion even more die, inasmuch as it is against nature. Is it not, then, necessary that the physician be a superintendent in the matter? But to avoid the trouble of being everywhere at once (even though most often there is not a lot for the midwives to do), it suffices that the women who practice this profession be instructed by physicians and know the reasons for what they do. Indeed, in a well-governed realm, physicians need to teach midwives the anatomy of the organs harboring the child, those that furnish it a passageway, and those that help push it out, so midwives can have a working knowledge of the true method of proceeding in their work.

Otherwise, they go about it blindly like empirics, without knowing what they are doing. Because of this ignorance, most of these women become arrogant and presumptuous, even extremely so if they have ever served some great lady or have been sent for from afar. This causes them to become haughty. Then, if a physician tells or shows them something, they will laugh at it or dismiss it. Terence spoke well in saying there is nothing more foul and unjust than an ignorant man, for he only finds value in what he does himself.[2]

I once visited with the late Monsieur Rondelet[3] a sick woman who was complaining greatly about the suffocation of her womb. We met there, on one occasion among many, an old matron who pushed us out and, as we were entering the room, said that the woman was not of our province, that she was pregnant, and that that was none of our business. As if we were not able to distinguish pregnancy from an unnatural condition, or as if the pregnant woman, who was also ill, could be exempt from our treatment. As it turned out, the woman was not even pregnant, and for three months the old matron had been there eating and drinking at the woman's expense for no reason. Oh, what folly! What temerity!

What disgusts me is how these women share among themselves a few small remedies, which, after all, are not even of their own invention but were taken at some time or other from physicians and later passed around among themselves. For women have never invented a single remedy; they all come from our domain or from that of our predecessors. They are very ignorant to think we do not know about these remedies and to think they know more about them than we do. But these good ladies admit openly the truth when they call upon us for help so often, unable to manage affairs on their own. For if we are able to do what is more difficult, do we not also know what is easier and more ordinary, like the alphabet? It is a little ridiculous to tell somebody who knows how to read and write that he does not know his letters!

CHAPTER IV
ON LEAVING A GOODLY LENGTH ON BOYS AND NOT ON GIRLS,[1] AND HOW TO CARE FOR THE NAVEL STRING; AND WHETHER GIRLS' NAVEL STRINGS HELP ATTRACT FUTURE LOVERS

Man is no sooner born than he undergoes surgery: the cutting of the navel string, which midwives remove after tying it off close to the belly, where the navel will be. Concerned for the preservation of the human species, midwives usually advise and kindly require matrons (when it is a son), to leave him a goodly length. For they think that the virile member will use it as a pattern, and that it will become longer if the string still hanging from the navel remains rather long.

But they are mistaken and have poorly retained what some ancient physicians might have on occasion advised midwives: that is, when they come to tying off the navel string of a boy, they should leave it very loose, without pulling on it. If they tie it off close to the belly, the bladder, which is supported by it, is drawn more toward the interior, and the virile member is thus shortened because the tube common to urine and semen is connected to the neck of the bladder. It is therefore important, for the length of the virile member, not to tie off the navel string too close to the belly and not to leave a lot of it hanging. For that serves no purpose. On the contrary, for girls it is good that it be drawn out and tied close to the belly, so that the womb, which is attached to the bladder, will, by being drawn in, cause the neck of the bladder to be narrowed and lengthened. And there is the secret.

We also advise that the navel string be tied off tightly, as it indeed should be. For when it is not tied off properly, some children die from losing all their blood through it. From just such a danger my wife, Louise de Guichard, suffered, as her mother recounts. Women who were present pronounced that she would never have good coloring in her face because she lost so much blood. But that prediction did not amount to anything.

I have another warning concerning health that must not be scorned as it commonly is. It is about the portion left hanging, which dies off little by little and finally falls off from gangrene, or, rather from sphacelation. Midwives usually lay it down against the naked skin of the child's belly. But this causes the poor child to feel sharp pains and aching in the belly. It cries night and day, without people knowing what is hurting it, and a thousand things are blamed that are not at fault. For example, in the region of Agen they blame it on bristles (that is, hairs like those on pigs or horses) that are in the child's belly (so they say) and give it sharp pains. So women lather the child, especially its belly, applying fomentations made with gentle lye from vine sprigs and a handful of straw ash. While rubbing the body of the child, the pieces of this straw ash stick out between their fingers, which they show to the people looking on and say are the bristles coming out of the child's body. And this is how the pain goes away, but it is actually the heat of these fomentations that drives out of the child's belly the coldness therein causing the pain (as with colic),[2] and not because there were bristles in there—as can truly be the case sometimes, however, in children's spines, an illness we shall discuss (God willing) in the fifth chapter of the eighteenth book.[3]

What hurts the child's stomach, then, is the string hanging from the navel, because of its coldness (a result of its mortification). For since the veins and arteries have been tied off, the natural heat is extinguished little by little until the outside part of it is completely dead and black. It is then extremely cold and rests on the child's belly like a piece of ice. We must, therefore, not be surprised if it cries and complains. In order to avoid and prevent this condition (having compassion on the poor little children, who cannot explain what is wrong), I order and advise that this part that is left hanging be from beginning to end carefully wrapped in cotton or soft linen cloth so that it cannot touch the naked belly. I find that this makes the child remain more calm. And so this is a sure sign (besides the apparent reason mentioned above) that it is the utter coldness of this hanging scab that causes the pain.

In some regions the women carefully keep the remains of their girls' navel strings in order to secure lovers when it comes time to marry them off. It is because they believe that if the man that appeals to them is given some of this navel string, reduced to a powder, in something he eats or drinks, he will fall violently in love with the girl, and there will be nothing left but to make up the marriage contract. I hold this to be foolishness and a most flagrant error, as I do most of what is said about other love potions, called in Greek *philtres*, attributed to witches and old whores for the purpose of pestering men with their love.

But I do think that there is a hidden allegory in such a belief, and it is (perhaps) that if men become so familiar with easy and pliant girls as to be able to touch and join their navels, then they do attract them with their navels. They thus make the conjunction of the Platonic androgyne, and through this union many are caught, sometimes to their detriment. This is how girls' navels—not dead ones, but live ones, with which men are put in appetite by being enticed—inflame and abase men, unless reason governs and dominates them. This is how it happens that they so often yield to the vile parts of their nature.

CHAPTER V
WHETHER IT IS TRUE THAT ONE CAN TELL FROM THE KNOTS IN THE CORD OF THE AFTERBIRTH HOW MANY CHILDREN THE WOMAN WHO JUST DELIVERED WILL HAVE

This notion can be attributed to Avicenna or to Rhazes,[1] who wrote that the way of knowing how many future children a woman who is delivering will have is to note and observe the umbilical vein, which is like a cord attaching the child to its afterbirth. As many knots, wrinkles, or kinks as there are in this cord, that is how many children the woman will have; and if there are no knots, she will have no more children. Also, if there is a lot of space between these knots, the woman, too, will have great intervals between her pregnancies; and if the distance is small, there will be scarcely any interval at all. Furthermore, if the knots are black or red, she will have that many males; and if they are white, daughters. Master Antonio Guainiero dares to say in his *Practica*, in the thirty-first chapter on afflictions of the womb, that in his day he found through experience that all of this was true.[2]

This is why one must not be surprised if laymen share this belief held by such serious authors, philosophers, and physicians. It would, therefore, seem wrong for us, if this is an error, to attribute it to laymen. I respond to this by saying that I want to turn laymen from their errors when they go astray in matters of medicine and knowledge of natural phenomena, whence many mistakes may spring. I also know very well and concede that most popular errors in matters of medicine and health are due to physicians' statements, either incorrectly understood or incorrectly retained.

It is also possible for there to have been false and erroneous doctrine, something we are quite familiar with and refute every day in our works and lectures. I am treating here only the most common errors, which are on the level of, and able to be grasped by, laymen. One such subject was treated earlier: that of the matrons and midwives who wanted to be considered prophetesses and played the pompous ones to perfection.[3] Because they use no logic or reasoning, what they have once understood and accepted as true, actual, and certain will be retained by them forever, like an oil spot. And in order to confirm them even more in their opinion, they have only to hear that it was said by one of the ancients, or that it is from the past, and right away the proposition will be completely approved, verified, and authorized. If they are told something better, either in being corrected or taught, they will take no account of it if it does not align with some other piece of their recollection. We should not be at all surprised at this, for there are plenty of men in the profession of letters who are just as stupid as this, even in their special subject.

I return to my point: what basis can there be for saying these knots predict how many children the woman will have?[4] I do not wish to present the objection that she could die because of some misfortune just a few months later, or that she could be so poorly cared for during her delivery that she will be sterile from then on and, consequently, will not have as many children as the knots had promised.[5] Such objections would be frivolous, especially since one must always suppose that there is no hindrance, such as her husband's death in the meantime, her unwillingness to remarry, and her living chastely in widowhood; the prediction would not prove false because of such things. For it is presumed that she will continue her course and perform the required acts; it suffices that she be apt and able to do what the knots promise.

But there is not the slightest semblance of truth in this observation, especially since the situation, number, and color of these knots is subject to the chance combination of the different elements, some arranged this way, some another. All the meaning these knots can have is the conjecture, in my opinion, that the great number of knots or twists that are next to each other, and red or blackish in color, might be a sign that the woman's womb is strong and of good constitution, good heat, and low moisture. For that which is thus knotted is also stronger, as we say concerning incisions in the long, straight muscle of the epigastrium,[6] and the color red is a sign of vigor. Thus, one could say that when there are many knots in the umbilical cord, the womb that formed them is healthy and will be able to produce many children. Not that one can guess the number, for the womb will be able to make more or less children than there are knots. One can likewise say that it will have them quickly, one after the other, without waiting, because of fecundity, and that there will be more males than females because such is the condition of a well-tempered womb. And this is all that the knots, whether in great number, close to one another, or of a red or blackish color, are able to predict.

CHAPTER VI
CONCERNING CHILDREN WHO ARE BORN "CLOTHED";[1] WHETHER THEY ARE MORE FORTUNATE THAN THE OTHERS, AND WHETHER THE CHILD'S BIGGIN PRESERVES FROM HARM THOSE WHO WEAR PART OF IT

This saying is even more inept than the previous one, if it is not taken in a mystical and hidden sense to mean something else that is not being said, as I shall interpret it. The child has at birth a tunic, or very thin membrane, that covers and surrounds it very tightly, just as the shroud does a dead body. In Greek it is called *amnion*, meaning biggin, thus named because of its thinness and fragility.[2] Around it there is another fleshy skin called the chorion, and the secundine, which is the couch, or afterbirth, to which the amniotic skin usually remains attached. The child sloughs it off completely, coming into the world entirely naked out of the womb, which is unclean, filthy, and foul,[3] located between the rectum and the bladder (whence the child is said to be lodged between piss and shit). And so the proverb of the local women in Languedoc is true indeed: *Entre la merde & lou pis, se nourris lou bel fis.*[4]

Sometimes it comes out wearing its tunic, like a shirt, which rarely covers its entire body and scarcely ever goes past the shoulders, sometimes only covering the face. This is taken as a good sign, and people say it will be fortunate because it was born clothed. Is this not an allegory for those who are born of rich and opulent parents, the kind that have nothing but their pleasures or honors to see to, unconstrained in any way by necessity? It is commonly said of such people that they are fortunate and born completely clothed, that is, with a lot of wealth acquired from their parents. The others, who are poor from birth, are truly born completely naked.

This is how I would like to interpret it. For there is no reason why the biggin should bring good fortune to those who keep it. It is by chance that this occurs, when the child is hardly impeded upon coming out. Other children, because of the considerable movement they have, shed it completely. But we could say that children who keep it are by nature gentle, sad, and peaceful, causing them to be of a greater modesty and thus making people cherish and love them, which would bring them great favor, fortune, and honor. On the other hand, we also say "Fortune favors the brave"; and brave people, being very active, could well have left their biggin behind. So there is no solid foundation for this prediction.

There is even less foundation in saying that such a biggin, or piece of it, keeps the person wearing it from peril and danger. It is true that if he is thrown from a horse and breaks his legs, the pieces will be found in his boots, if he happens to be wearing any. What stupidity! It is like the little spells some people cast so as not to drown, get burned, or break their necks even though they jump into a deep river, into a roaring fire, or from a high place. There are some who say they know how to conjure firearms so that you will not be struck or wounded by them, and that they know how to cast a spell on a man so he will not be wounded in battle even if surrounded by a hundred of the enemy. Go ahead to the besieging of a city, protected only by these spells, or by these biggins, and you will see whether this biggin guard, spell guard, or other bravado will save you. I think you would soon find yourself trussed up like a ninny. I would much rather have Grimache's recipe for a day of battle:[5]

> See to it that, by design,
> You approach the battlefield by no more
> Than thirty leagues at the closest,
> Or that you arrive only too late,
> When all the blows have already been dealt.

Here we have more good sense than rhyme, but in the others there is neither rhyme nor reason. I am willing to grant that there are spells that cure fevers, stop bleeding, and achieve other great effects because of the belief one has in them, in conjunction with a strong imagination. But as for stopping accidents from happening and repelling bad things that come from the outside, that is another matter.

CHAPTER VII
CONCERNING THE HARPIES THAT PEOPLE SAY FLY UP AND STICK TO THE BED CURTAINS

To designate a very unusual and monstrous creature with claws, people use the word *harpy*. It is an allusion to those fictional Harpies of the poets, mentioned by Virgil in the third book of the *Aeneid*, in which he has three of them, describing them as having the faces of women, clawed hands, and bellies full of filth with which they infected every food they touched when unable to steal and carry it off.[1] They were monstrous and rapacious birds (as the name *Harpy* suggests) sent by the gods as a punishment to Phineus, King of Arcadia. They stole his food and, after blinding him, befouled his table with great amounts of reeking filth. And this was because he cruelly put out the eyes of the children of his first wife, having since married her stepmother. Some time afterward, the Harpies were chased away from this miserable king by the brothers Calaïs and Zetes, who also flew like birds.[2]

Monsieur Ludovico Ariosto in his *Orlando furioso* gives a pleasant account of this fable, furnishing the following version.[3] The emperor Senabo, or Preteian (as he is called more specifically), of Ethiopia was so fearless and arrogant that he wanted to do combat with God in the place he was told the earthly paradise once was. For this outrage he was punished by the death of his people, up to one thousand of them, and by being blinded. Besides this there were sent from hell seven Harpies with the pale and dead faces of women weakened and dried by extended hunger, more horrible to look at than death itself. They had huge, deformed, and ugly wings, rapacious hands, crooked and twisted fingernails, huge and foul-smelling bellies, long serpentlike tails that could go into coils and knots. These Harpies no sooner smelled the food that the emperor (king of Nubia, where he lived) was being served, than they were there, beasts that they were, turning over all the dishes and stealing the food. What they were unable to swallow, they shat upon with such a foul-smelling excrement that no one could come near it. And so this poor man was dying of hunger, until Astolpho, with his horn and mounted upon his hippogriff, delivered him. Now all these are fables and poetic inventions in which, nevertheless, there is gracious instruction subtly hidden.

But let us come back to our point. It is certain that women form and deliver monsters, called in French *amas*. They are like lumps of flesh that have no shape or distinctive form and are engendered in the womb, sometimes because of corrupted semen, which is incapable of forming a child. Because of the menstrual blood flowing around it or drawn to it, such a fleshy mass filled with sinewy filaments is formed. In other cases it is the result of the woman alone, who has corruption within herself, for she has both the blood and the seed to produce such a mass.

Sometimes these masses are the only things there, and the woman thinks she is pregnant; sometimes they accompany the child, which they will often harm by taking its nourishment, occasionally causing a miscarriage. For the child does not have enough room or food to reach maturity. This is not something rare, not as rare as what is written about certain animals that sometimes incubate in the womb from corrupt matter which has been retained, just as in the stomach and bowels large fat worms incubate strangely. Some have written of a scorpion that was found to have been engendered in a man's brain. In this same vein, some say they have seen strange creatures come out of the womb alive and moving, looking like toads and other ugly animals.

Niccoli of Florence compares them to screech owls or horned owls and harpies, and says that in certain regions they are called wild-creatures, or evil-creatures, and that sometimes they bite the child and kill it, and that in Pisa and even more so in Apulia (in the kingdom of Naples), the women are very likely to have them because of the poor food. He furthermore cites a man whose wife brought forth in one day nine separate and deformed masses of flesh, none of which resembled any of the others, and each one of which weighed from four to eight ounces. These are indeed moles, or mooncalves, which lawyers also call harpies. They also call them brothers of the Lombards because the women of Lombardy are particularly subject to having them (as Gordon has written)[4] because of their poor food, fruit, and plants, preferring to be better dressed than nourished. This is why people in France say that the female must be well dressed and ill fed. To which they add: and well beaten! Males, on the contrary, must be better fed than dressed.

The Sire d'Aubigné,[5] squire of the king of Navarre, told me that when he was a schoolboy staying in Geneva in 1565 in the house of Monsieur Philibert Sarazin,[6] a most learned physician, two Italian women, one the wife of a mender and the other a young gentlewoman, brought forth in the same month monstrous burdens. The mender's wife had one that resembled a rat without a tail. The young gentlewoman's was as big as a cat. Both of them were of a black and viscous substance. Upon freeing themselves from the womb, these monsters flew up and firmly attached themselves to the partition of the side of the bed, higher than the canopy. People able to be trusted tell of a most decent woman from Châtellerault who had such a monster. It escaped from the hands of the midwife and ran all about in the bedroom like a hobgoblin, black and of a strange shape. It was finally caught and smothered by the midwife. Yet it is said that these monsters must not be harmed, that the woman is to remain in bed for nine days at the least, that she must finish her lying-in as if she had had a child, or else she will also die. But no such harm came to this woman, who has since had beautiful children.[7]

This is what is related concerning such things. Let us now see what is to be believed. It is certainly true that women engender and expel from their wombs some time after their flowers have stopped flowing (thinking surely that they were pregnant) lumps of sinewy flesh,[8] which can be compared to this and to that because of some resemblance they have to certain things, just as is said of clouds, that one looks like a horse, another an inkhorn, another a hare, another a bird. But harpies are nothing like all these, and their tissue has only vegetative life, like a plant, simply, without any movement of its own, or any feeling.

Thus, it never was an animal, not even a reptile, or some other less perfect form of life. This is why it is utter foolishness to believe that some of them fly about like harpies and suddenly go attaching themselves to the curtains of a bed prepared for a woman about to deliver. I did not retain very well what some Neapolitans told me about this, or what happened in the end, and what it all meant. But one is not going to be damned if one does not believe it. It is commonly said when someone tells a strange and rare story (in other words, an unbelievable one): "If I hadn't seen it, I wouldn't believe it." By this phrase and manner of speaking, one excuses and dispenses those who have not seen it from believing any of it, even if one convinces them of it. For by saying "If I hadn't seen it, I wouldn't believe it," it is as if one were saying "I advise those who haven't seen it not to believe it."

This is what we can say of these monstrous mooncalves called harpies, said to fly like birds. It is unlikely that our lawyers, who have named them harpies, think that they are real animals, and even less that they have wings to fly with, but simply call them thus as a means of comparison to something very deformed. Likewise, the Harpies we described according to the poets are not at all real, but made up. Regarding the expression "brothers of the Lombards," it is due to the fact that the wives of the Lombards (a most odious nation long ago) were subject to having such monsters. Because this mass is taken for a monstrous child, it is called a brother of the others that are perfect and normal. For they are conceived in the same womb and nourished on the same blood. This is why they can be called, as a detraction on the part of people who despise them, uterine brothers.

CHAPTER VIII
WHETHER IT IS TRUE THAT A WOMAN DELIVERING UNDER A FULL MOON WILL AFTERWARD HAVE A SON, AND IF UNDER A NEW MOON, A DAUGHTER

Some people hold this belief, maintaining that if a woman brings forth a child under a full moon, when she next delivers, she will have a son. They say they have observed this, and that it never fails. I do not contradict this and grant willingly that they have never seen it happen any other way in the many women (even up to a thousand) they have observed. But I claim that it does not happen like this to all women, and especially not to one I was able to observe, who had several children. For I do not stop after cases of two or three children, but for brevity's sake I shall be content to cite the children that God has given my late father, Monsieur le Chevalier Joubert,[1] and my mother, the gentlewoman Catherine de Genas (still living), twenty in number, all of the same marriage.

Jane was first, born in the year 1519 on the sixth of July at seven in the morning, under a new moon.[2] After, there came Marguerite in the year 1520 on the twentieth of July at six in the morning, under a new moon. Susanne succeeded her, born in the year 1521 on the ninth of July at one in the afternoon, under an old moon.[3] Fleurie followed in the year 1522 on the twentieth of July at seven in the morning, under an old moon. Another Jane was born in the year 1523 on the twenty-fourth of August at nine in the morning, under a full moon.

After all these girls there came two sons: François, who was born in the year 1524 on the fifteenth of November at midnight, under an old moon; the other, named Guillaume, was born in the year 1526 on the sixteenth of January at two in the morning, under a new moon. After, there came two daughters: Magdaleine in the year 1527 on the twenty-sixth of January in the morning, under an old moon;[4] and Catherine in the year 1528 on the seventh of May at three in the morning, under an old moon.

I came next, born in the year 1529 on the sixteenth of December at nine in the morning, under an old moon. Then came Antoine in the year 1531 on the eleventh of January at six in the morning, under an old moon. Ysabeau succeeded him in the year 1532 on the fourteenth of December at seven in the evening, under an old moon. After, Anne came in the year 1534 on the seventeenth of June at six in the evening, under a new moon. Next came the twins, Louise and Justine, who were born in the year 1535 on the seventeenth of July at eight in the morning, under a full moon.

After, there happened along a son, named Antoine II, in the year 1536 on the twentieth of October at seven in the morning, under a new moon. Following him, there also happened along a daughter, named Dauphine, in the year 1537 on the eighth of November at five in the morning, under a new moon. Then there was born a daughter, named Françoise, in the year 1538 on the fifteenth of December, one hour after midnight, under a full moon. There followed a son, Claude, in the year 1540 on the ninth of June at six in the morning, under a new moon. Then came another son, named Felix, the last child, who was born in the year 1541 on the fourth of October at eleven in the morning, under a full moon.

From this genealogy, transcribed directly from my late father's register (except for the particular moon, which I based upon the ephemerides of the aforementioned years),[5] it can easily be understood that there is no certitude in such a statement.[6] I observed it even better in the children God has given me, up to the present, from Louise de Guichard, my wife.[7] Isaac was born on the third of March 1565, under an old moon; Susanne, on the thirteenth of the same month in the year 1567, under an old moon; Anne, on the same day one year later, under a new moon; Marie, on the twenty-ninth of July 1571, under an old moon; Cyprien, on the fourth of August 1574, under a new moon; and Henry, on the sixth of June 1577, under a full moon.[8] From this it is seen that this saying is true for Marie and Cyprien, but false for Susanne, Anne, and Henry.[9]

Just as false,[10] or even more so, is what our women say about the first child (whether it is a male or a female): that if it is born under an old moon, the mother will henceforward have nine daughters in a row. In this there is less of a foundation than in the preceding error, and experience is to the contrary. For the first of my children was a son, followed by three daughters, and then two sons. It is true that my wife was most afraid that she would prove this saying true when she saw she was continuing with daughters (on up to three), even though I demonstrated to her that the belief was ill founded.

There is another error regarding this saying. Some people maintain that if the moon turns (that is to say, becomes new) within nine days after the birth, the woman will change the sex of her next child, for example, if she has a son now, the next child will be a girl, and vice versa. And if the moon does not turn within this nine-day period, the woman will continue in the same way. But we have shown elsewhere that the moon,[11] even though in such and such a position at the time of delivery, does not predispose the womb for the future to any important extent, as it is well able to do at the time of conception.

CHAPTER IX
CONCERNING ALMOND OIL AND CANE SUGAR, WHICH SOME WOMEN DRINK AS SOON AS THEY DELIVER; AND CONCERNING FOOD THAT IS GIVEN THEM INAPPROPRIATELY

In Languedoc and in some other regions, it is very common to give women three spoonfuls of almond oil with a little cane sugar immediately after delivery. In other places people have quick recourse to hippocras.[1] Women from Normandy take a warm drink made with milk, cinnamon, and sugar. Others have capon broth or chicken consommé; still others, the yellow of one or two eggs with a little sugar but no salt, because of the feared immediate drying it causes. Others have different foods, according to availability and personal means.

On all of these it is necessary to give some warnings, as we shall soon do, after having discussed almond oil. I think these women have adopted this practice for two reasons, mainly. The first is that many women are in labor a long time, and because of the fierce pain they scream at the top of their lungs for long periods of time. This is not to be reproached, for it helps with the delivery considerably because of the great pressure exerted on the muscles of the lower belly and on those of the chest and diaphragm. Thus, the womb is pressured, pushed upon, and constrained so that through such means it is emptied and discharges its contents more easily. One does just as well without screaming by holding one's breath and straining as when one wishes to void a tightly constipated belly.

But it is important for the woman in labor to use these remedies appropriately, reserving them for forcing the child and the womb, and not screaming or straining with every pain she feels. For it could happen that when the need arises she will not have the strength to use such means (which help the child and the womb considerably) because she is greatly exhausted and worn out from straining and screaming.[2] Now, from all this it often comes about that the woman in labor has great dryness in her throat and a bitterness that makes her hoarse, for which the oil mentioned above (along with cane sugar) is very good in relieving, moistening, and soothing the throat, as well as restoring the voice to its full level.

Women can also be of another opinion: that this oil prevents pain or causes them to have less of it. For in such instances some of them drink a dish of olive oil or walnut oil. It is true that these oils soothe the belly and dispel pain in the parts they reach, such as the bowels, for such oils are very soothing and relaxing, especially very mild olive oil and almond oil. But they do not reach the womb nor the blood vessels, which then overflow and spill forth the superfluous blood that had been retained for the child. It is at this point that pain begins, when the thick and heavy blood, like the dregs and deposit in wine, builds up everywhere and flows through the veins and arteries to the womb, which it penetrates with difficulty and with much violence, because it is being repelled as so much useless matter.

These are the principal causes of such pains. There can also be some coldness because of the cold air that enters the womb after the child exits, and even more of it if the woman is not well cared for and is left in a draft, or if one forgets to put the very warm afterbirth on her belly immediately, or a slight amount of pressure on it with her legs crossed, in order to keep her from getting her womb chilled, which is greatly to be feared.[3] Since these are the causes of such aches and pains, of what use is oil, which does not enter the womb or blood vessels and does not even reach them? In fact, it goes right through the bowels to exit in the fundament.

I reply by saying that, upon arriving in the large bowels, named the colon and the rectum,[4] it acts upon them like a fomentation applied internally and very locally, such that this oil obviously allays and soothes the pain and causes the superfluities to be voided more easily. For the oil is in the bowels, which are so adjacent to the womb and the vessels mentioned above that these organs are also fomented.

Now let us see whether it is good to give a woman something to eat as soon as she has delivered. It seems to me that one is quite mistaken when one does so with all women, indifferently and without exception. For perhaps one woman had dined or supped heavily just before delivering. What need does she have of a good soup, or of fresh eggs or other food, since she has enough food in her stomach, still raw and undigested? It is not good to put raw on top of raw and thus overburden the stomach, which, rather than being fortified, will become weakened by it, as will the entire body. As for giving her a little to drink and to eat (as one does otherwise, without having delivered, two or three hours after a meal), there is no harm in this, especially since she deserves a drink of something after her straining and screaming. But as for nourishing her so inappropriately and without any need, I cannot give my approval.

In order to avoid fevers and other dangerous complications, it is necessary from then on to begin nourishing her more lightly, like a person who has been wounded. Indeed, one could not more aptly compare a woman who has just delivered than to a person who has just been severely wounded. There will still be the difference that, in the case of the wounded person, bleeding is stopped immediately because the blood is good; whereas in the case of the woman this cannot be done because this blood is not worth anything, at least for the most part. She must be nourished little by little until the complications of pain, fever, and other common ailments disappear, and until the woman is well purged, which can be accomplished in eight days if she is well cared for. Then, like a person recovering from an illness, she must begin to be better nourished. And within another eight days she can be well enough (if she is of a sound constitution and healthy) to wash and bathe the next week and to go out of the house (if it is the custom of the region, for otherwise she would be reprimanded by the other women)[5] on the twenty-first day. For the twentieth is the outer limit of acute illnesses,[6] without a relapse or a foreshortening, in accordance with orders from physicians.

But whence came the custom of preparing and presenting these foods as soon as the woman has delivered? It is very ancient, as I see it, and has been observed since men have become more continent, only making love to their wives in the morning, after some sleep and much rest. Because of this the children also were stronger, in line with what I said in the second book, chapter seven. And so it came about most often that women were delivering at about the same time of day, having completed the cycle of maturity for their fruit. Hence, a bouillon or some other food was most appropriate, since the woman, having started her labor early in the morning, had earned a meal by the time she finished the task. Now that people are more given over to pleasure and carnal voluptuousness, they do their business at all hours of the day and night, and most often after a meal—quite inappropriately—as I have also pointed out in the chapter mentioned above. And so, because of this, women today deliver at all hours of the day and night. But this is not to say, however, that they should therefore be given bouillon or other food at all hours, if there is no need or necessity.

CHAPTER X

THAT SOME GIVE EXCESSIVE NOURISHMENT TO WOMEN WHO HAVE JUST DELIVERED, CLAIMING THAT THE WOMB IS EMPTY AND THAT IT MUST BE FILLED

If one has begun poorly, one does worse to continue. I am not speaking of nourishing, but of overfilling and stuffing to the point of bursting women who have just delivered, as if one wished to make sausages of their bellies. Women justify this practice by saying that the womb is empty and that it must be filled. It is a principle of natural philosophy and is very normal that nature abhors a vacuum and cannot abide one anywhere.[1] But the womb, which continues emptying itself for several days after childbirth until there is no longer anything superfluous left over, tightens and closes up so that it never has any empty space needing repletion. And if it did need repletion, it would not be food that it required, nor blood made from food, but sperm alone, which is its delight and the thing most desired.

But I am assured that decent women will not let it have any before their lying-in is duly celebrated.[2] Thus, there is no reason for such ample nourishment of women who have just delivered, especially during the first few days. This would only serve to make matters worse, maintaining or elevating a fever and causing more pain in their breasts. It is necessary to go about it gently, as with people who are wounded, as we said in the preceding chapter. Still in all, being mindful of evacuation (even if such was necessary), they must be nourished after the first seven or eight days, and even more so if they wish to nurse, as indeed duty requires them, which I shall sufficiently demonstrate in the beginning of the next book.

CHAPTER XI
WHETHER IT IS TRUE THAT A WOMAN WHO HAS JUST DELIVERED IS ABLE TO PISS MILK

Several people find it strange that our women commonly say: "She is pissing milk," as if it were something impossible or absurd. Yet, I have seen it happen frequently, not so often by itself but as a result of the application of remedies for drying up the milk glands. For some of these remedies are so strong that they repel and suppress the milk already formed inside and force it into the vena cava. If it is not milk, it is at the very least pituitous blood (suitable for the production of milk), somewhat whitened, which[1] returns to the large vessels and, from there, is drawn off by the pumping vein[2] and then voided in the urine, which therefore becomes white.

Sometimes it is on account of the spontaneous return of this substance, without any repulsion, as is the case when the parturient woman does not nurse. For the substance that forms in the milk glands is maintained by frequent suckling, otherwise it would not keep being produced for very long.

But how can it be that milk carried with the blood in the large vessels is able to retain its color?[3] It is easy to see that this is feasible, because the foul matter from an aposteme[4] on the liver, the spleen, the lung, or other internal organ, depending on how much it is broken down, can be seen in the urine as white or red matter. If this matter does not change its color upon being mixed with the blood, then neither will the milk. The reason is rather obvious when we realize that we have in the organs of our bodies a secreting or separating faculty that can select the good from the bad among the substances previously combined and mixed. The gall bladder, for example, draws to itself the choleric portion of the blood, which is not visible to our sense of sight when in the blood, and the kidneys draw out the serosity, or water, of the blood and set it aside. The kidneys are also able to remove from the entire sum of blood, or sanguinary mass, the pituitous, already whitened, and half-milky portion that is rejected by the milk glands. Thus, what these people say is not absurd: that women piss milk.

CHAPTER XII
WHY IT IS THAT WITH THE FIRST CHILD ONE USUALLY HAS FEWER SHARP PAINS

In the ninth chapter of this book, we quite amply discussed the causes of the pains women have in childbirth. Here we must accept as solid conclusions what was demonstrated there: namely, that thick, heavy blood, like wine dregs, penetrates the womb with difficulty, chilling it and making it swell.[1]

Now, with the first child, the womb is not as loose as it will be with future ones because of expansion. This is why with later children it is more susceptible to taking in air and being injured by it. As for the blood, it keeps on getting heavier and thicker, which is why it is more difficult for it to flow out. There are even women who are not pregnant but who, on the verge of their menses, have terribly sharp pains in their bellies and aches in their loins, because their blood is very coarse and flows with difficulty.

To these reasons can be added the fact that the pain gets worse as it returns. This is because if an organ is once distressed and feels pain, when the pain returns it will be much more severe, for the organ is weaker than it was and thus more sensitive. This is why (in my opinion) one has less pain with the first child.

Simple people give another reason: that is the way God wishes it, so that the woman will not be temporarily disgusted with the idea of having children again.[2] But one can clearly see that even after the most painful births, they are just as ready, or even more so, to start all over again. Even if they were near dying, all the pain is forgotten, and the good ladies are once again more than willing. The moon has not yet completed its cycle, but they are ready to begin again. You would think they never suffered, so pliant, kind, and receptive are they to every proposition. Even though there comes to them from this encounter a great effusion of blood, they are so gracious that no sooner does the wound cease bleeding than the only memory they have is of their first loves. Oh, great bounteousness of the feminine sex! It always loves the more those who bring it so many afflictions—afflictions from which several of them sometimes die.

END OF THE FOURTH BOOK

THE FIFTH BOOK OF POPULAR ERRORS CONCERNING MILK AND THE CHILD'S NOURISHMENT

CHAPTER I
EXHORTATION TO ALL MOTHERS TO NURSE THEIR CHILDREN

Aulus Gellius relates a beautiful lecture delivered by the Athenian philosopher Favorinus, which I thought I would reproduce here as a preamble to my discussion.[1]

Aulus Gellius recounts that someone told Favorinus that the wife of one of his disciples had delivered a son. "Let us go and see the woman," he said, "and congratulate the father." For the disciple was of senatorial rank and from a most noble house. Favorinus, followed by his disciples, entered the house. Having embraced and congratulated the father, he sat down and asked how long the wife had been in labor and what troubles she had had. When he was told that the young woman, exhausted from her labor and lack of rest, was sleeping, he decided to stay longer and talk.

"I am sure," he said, "that she will nurse this son with her own milk." The mother of the young woman replied that she must be spared such a trial. The child would be given to nurses so as not to add the worrisome and difficult task of nursing to the pain the young mother suffered in bringing forth the child, for the poor mother was too young and fragile.

Then Favorinus said to her, "I pray you, my lady, to allow her to be the whole and complete mother of her son. Only an unnatural, imperfect, and halfhearted mother would have a child and then suddenly reject it and send it away. What sort of woman would nourish in her womb with her very blood a being she could not see and then not nurse with her milk one she could see, alive and expecting its mother to do her duty? Do you think nature gave women nipples on their breasts to serve as pretty little buttons for embellishment rather than for the nursing of their children? Are these women not prodigal when they work at stopping and drying up these most sacred fountains of the body, wellsprings of the human species (even to the detriment of their health, on account of the clotting and putrefaction of the milk), as if it disfigured the marks of their beauty?

"What difference is there between this folly and the madness of those women who, through certain wicked inventions, force themselves to abort, so that the smoothness and polished flatness of the belly might not be spoiled, so that it might not be stretched, filled, and split open by the weight of the burden and by the throes of childbirth? One ought to publicly decry, detest, and deplore those who put a man to death at his very beginning, when he is being formed and receiving life, when in the very hands of nature and in the process of being fashioned!

"And how close to this wickedness are those mothers who deprive their children, already complete and born, of the nourishment of their own blood, with which they have become familiar and accustomed! 'But there is no difference,' they say, 'so long as the child lives and is nourished with any kind of milk.' Why is it, then, that people answering thus (if they are so dull in comprehending nature's feelings) do not also think that there is no difference as to the body in which the child is conceived, or with what blood it is engendered? Indeed, before a marriage, people look very deeply into the man's and the woman's circumstances, their ancestry, their blood, and their conduct, so as to have the best lineage possible. But is not the blood that was then in the womb the same as that which is now in the breasts, only whitened with a large amount of spirits through natural heat? Does one not see in this process nature's obvious industry and providence? This blood acts as the body's liquid craftsman; after forming the child in the lower regions, it turns upward at the time of delivery and flows into the higher parts (namely, the milk glands) and there stations itself, ready to maintain the blossoming of life and to offer the newborn a food known and relished by him.

"Certainly it is not believed for no reason that, just as the sperm has the power to make children resemble their parents both in body and mind, milk has the virtue and property of doing likewise. This is seen not only in men but in animals as well. For if a kid is nursed by a ewe, or a lamb by a goat, it is certain that the coat of the former will be softer and the wool of the latter will be coarser. The same is true with trees and fruits of the earth: the quality of the soil and the water nourishing them does more for their growth (or stunting) than the nature of their seed when it was sown. And very often a beautiful tree, green and laden with fruit in one region, when transplanted into another, will weaken and shrink because of the humor of the place.

"What (mau-loubet),[2] then, is this practice of corrupting the wealth, the value, the mind, and the body of the newborn child, which had such a happy beginning, and vitiating everything with the borrowed and degenerate food of foreign milk? This can well be the case if the child's nurse is of a servile, mean, and slavish nature or of a barbarian race, or if she is wicked, ugly, lascivious, or a drunk. For people will most often take indifferently the first nurse they find with a lot of milk. Shall we, then, allow our decent and wellborn children to be infected with the pernicious contagion it will draw into its own soul and body from the spirits of such a wicked woman?

"It continually amazes us that the children of decent women do not resemble their parents, either in mind or in body. When the expert and most wise Virgil imitated these lines of Homer:

> Your father was never the knight of Peleus,
> Nor your mother Thetis. The blue and billowy sea
> Has engendered you, felon, with the tall rocks,
> For you have a wild soul in your flesh . . . [3]

he accused not only nature, or geniture, but also this wild and cruel nursing practice. He even added on his own:

> The tigers of Hyrcania were your nurses.

And this is because the nurse's spirits, carried into her milk, have great power and influence in imprinting certain behaviors and complexions in those who first drink them, even more than those coming from the blood and spirits of the father and the mother by way of their sperm. Moreover, who could forget or neglect the fact that mothers who thus abandon and turn out their children, giving them to others to be nursed, sever the bond and tie of love, by which nature joins fathers and mothers to their children, or at least loosen and weaken this link. For often, once the mother has removed from her sight the child she has turned over to be taken elsewhere, the ardent strength of maternal affection is extinguished little by little, and all the vehement concern and clamor she used to voice for the child is silenced. One forgets a son sent away to a nurse almost as much as one lost to death.

"Alas, conversely, the affection of the child, its love and intimacy, go entirely to the one nursing it, and because of this, it has no feeling or desire whatsoever for the mother, who engendered it. They are just like children who are exposed. Having erased and abolished completely from their minds the elements of natural devotion, the only feeling that children who are thus remotely nursed seem to have for their fathers and mothers is for the most part politeness and civility, not natural love."

There you have fairly closely what Favorinus said, to which I will add a few points and fair examples offered by Dom Antonio de Guevera in his *Dial of Princes*.[4] Then I shall present several possible misfortunes for every kind and rank of women who refuse to nurse their children.

Is it not a type of madness to scorn what one had strongly desired, awaited, and finally secured? One of a woman's greatest desires is to find herself pregnant and then graced with a beautiful birth. How can she suddenly be so inconstant and fickle that, scarcely does the child see the light of day, she rids herself of it, sending it away to be nursed by a strange woman? I would offer the example here, first of all, of the other animals, far more reasonable in this matter than these women in nourishing without help every single one of their little ones with their own milk (at least those that have milk, for birds feed theirs with what they find in the fields). But I know that one would reply immediately that these are only animals, and they have not the wherewithal to replace one another. A female animal would not want to nurse the fawn of another animal, and so each is forced to nourish its own. Women, on the contrary, as social animals and of a friendly nature, have commerce with one another, facilitated by some just recompense. To which I reply by saying that animals are of such an affectionate nature with their little ones that even if they could be thus replaced, they would never permit it, as is seen every day in the great fuss they make with those who wish to take their little ones away from them, either to have them nursed by another animal, or for some other reason. And when, I ask you, does one find animals to be more fierce? Is it not when they are nursing? Very often they could flee and get away by running from the hunter who wants to catch them. But if this entails abandoning their little ones, they would rather be torn to pieces than leave them behind and lose them. Thus (as Plato says on this account),[5] children never love their fathers and mothers so much as when the fathers have carried them in their arms and the mothers have nursed them at their breasts.

Now, that alimentation has a great deal to do with the complexion of the body has been sufficiently pointed out above in the case of the food of a kid and a lamb. For the lamb that has been nursed by a goat will not only have a coarser wool but will also be more wild than it is by nature. I demonstrated this in more detail in the defense of my doctorate in Montpellier (which figures in the first decade of my *Paradoxes*),[6] where one can see the power nurture and education have in changing behavior and nature (understanding by nurture that which goes beyond nature, not only discipline and instruction but also manner of living and quality of food).

If there is a women reading this who is the least bit reasonable and who wishes to be convinced of her duty, she can have explained to her by a man of letters what I have proved in the *Paradoxes*. For the others, who plug their ears to every good argument, no more discussion is necessary. "He who cares not to do good," as the proverb goes, "has enough preaching." Yet I will pursue this point further just in case I might win over and convert somebody. I write for good and virtuous women who only fail in their duty through ignorance. We are not concerned about foolish and wicked ones; they no more deserve to nurse children than to have them. For we would have to fear that if they nursed their children, the children would also become wicked, and the world would become still more corrupt and troubled by their pernicious kind. It is already evil enough to be conceived by an evil woman and nourished on her blood for nine months in her belly, without catching still more of her evil qualities by sucking them in with her milk. It is good, therefore, to take them away from these evil mothers as soon as they are born and give them to good and kind nurses, healthy in body and mind, in order to blot out with a better sap the bad constitution imprinted in their bodies from the mothers' bad humors, which could cause similar behavior. Hence, one transplants trees and other plants into better soil to improve them, one washes drugs in several good solutions to remove evil natural qualities and to replace them with the good ones required for man's health. This is why it is said that Alcibiades, a native of Athens, was very hardy and valiant—uncharacteristic of the Athenians—because (as Plato says) he had been nursed by a woman of Sparta.[7] The people of Sparta were of a virile and courageous nature; the Athenians, on the contrary, were effeminate (whence Diogenes once said, coming to Athens from Sparta, that he was leaving the men and was on his way to the women).

These are important points that proper ladies ought well to consider and weigh in their scales of justice. They should, moreover, suspect that when prudent men consent to their wives' unwillingness to nurse their children, it might be because of the poor opinion (or certain knowledge) these men have regarding the bad conduct and wicked nature of their wives.[8] As for myself, I am of the belief that if my wife were sullied with any vice I knew of, I would in no way permit her to nurse our children, and so should everyone do. And women ought to consider themselves rebuked and held in bad opinion with respect to their husbands if they are not asked by them to nurse their children. For a husband who does not solicit his wife to nurse (when she is of a healthy constitution and able to) does her as much dishonor as if he were to say publicly: "My wife is not wellborn or well brought up; I don't want my children to take after her." Good God! What an outrage, if only women could recognize it!

Since, then, this practice is only for the wise, why do not all virtuous women make manifest their wisdom in this matter and leave the ranks of the foolish ones? I think, also, that if they knew what pleasure there is in nursing one's child—a pleasure their nurses enjoy—they would sooner lease themselves out for nursing other people's children than abandon their own. And why is it that nurses are ordinarily so fond of and thus attached to children who are otherwise strangers to them, if it is not because of the extreme pleasure they take in it? This pleasure is beyond comparison, far greater than all the sufferings children cause, gently smoothing over the vexations of the servitude and bad times one undergoes because of them.

I ask you to consider a little the pleasure that the child gives when it laughs, how it half closes its eyes; when it wants to cry, how it sticks out its lip; when it wants to speak, how it makes gestures and signs with its little fingers, how it stutters with such grace and repeats a few words, imitating the language it is learning; when it wants to walk, how it teeters on its little feet. Is there any enjoyment comparable to that given by a child who lulls and flatters its nurse while suckling, when with one hand it uncovers and plays with the nurse's free breast, and with the other takes hold of a strand of her hair, or her collar, fingering it; when it starts kicking at those who want to stop it from doing these things and at the same time gives out a thousand gracious little laughs and glances toward its nurse? What pleasure there is in seeing how it becomes spiteful and angry over nothing, upset over a pin, throwing itself down, beating and pummeling those who want to appease it, pick it up, or carry it away. What pleasure in seeing it reject the gold, silver, rings, and jewels one offers it in reconciliation, only to be suddenly won over with an apple or a piece of straw! What pleasure it is to hear the follies of little children, to see their games, to hear what they reply to requests, to hear their questions and the childish reasoning they give, the foolish things they say, and the statements they come up with from God knows where! Indeed, it is true when they say that where there are children there is no need of jokers or fools.

Is there not immense pleasure in seeing them play with dogs and cats, running after them? Or in watching them knead mud and make houses out of it, or ovens; or play the harquebusier, the lancer, the pikeman? Or watching them beat the drums, curtsy or bow, and imitate the wise man? Or see them weep over a sparrow that the cat has taken from them, or over the birds that fly and that they cannot have, or over a walnut that they lost, and other worthless things?[9]

Is there not pleasure and enjoyment when they do not want to leave their mother, or their nurse, and in order to make them go to another person some present or cajolement must be offered, and the mother or nurse must cleverly disappear? And when they will not allow their nurse to hold another child in their presence or to nurse it? When they begin to defend her if someone bothers her or pretends to beat her, how they are the first to weep and to become violent in order to vindicate the outrage? This great love, bordering on jealousy, is so pleasant and enjoyable that it thrills the heart of the nurse if she is of a good, human, and gracious nature. In fact, she will not love her own children any more than the outsider she is nursing.

And what must it be like when the mother herself is the nurse? If you take pleasure in what another has accomplished, such as a book or a painting, or some other fashioned object, how much more will you take from that which comes out of your mind? Without a doubt, the love and the pleasure are double with mothers who nurse their children. For, on the contrary, God quite often allows children to love their nurses more than their mothers. Of this we read several examples, which I shall furnish as succinctly as I am able.

Cornelius Scipio, surnamed Asianus,[10] having condemned to death ten of his most valiant captains for violating the temple of the Vestals, scorned the intercession of the most eminent people in Rome, who had begged him to relax the law and pardon the captains. He did not even take account of the importunate prayers made to him by his twin brother, Scipio, surnamed Africanus. Yet he was won over by the earnest prayers of one of his foster sisters. And when his brother reproached him for his discourteousness, he replied that he considered the woman who had nursed him without any natural obligation more of a mother than the one who had given birth to him. We read of two cruel tyrants—monsters of nature—the most wicked and awful who ever were, Nero, a Roman, and Antipater, a Greek. Steeped in other horrible cruelties, they refused to spare their mothers' lives, from which they had obtained their own. But nowhere do we read that these infamous villains, nor any other devilish tyrants, have ever hurt their nurses. The two Roman Gracchi, most valiant and famous captains, had an equally hardy and virtuous bastard brother.[11] When he returned from the Asian wars, in which he had been very successful, he happened upon his mother and his nurse. He first gave his nurse a gold belt, then his mother a silver ring. The mother was shamed by this treatment and reproached him for it, to which he responded that he was more beholden to his nurse. "For you, my mother," said he, "only carried me for nine months in your womb, at your ease, and only nourished me with your blood; and as soon as you saw me born, able then to rid yourself of me, you abandoned me. At that time my nurse received me lovingly, carried me in her arms, and nourished me with her milk for three years, a purely voluntary thing, and not out of some natural necessity, such as carrying a child in one's womb and nourishing it with one's blood. Thus, I feel more indebted to her than to you, as I wished to demonstrate through the difference between the presents."

These are beautiful examples that ought to spur decent and virtuous women on, encouraging them, even compelling them, to nurse their children, not allow a strange woman to have the best portion of their love—and the greatest pleasure children give. Several kingdoms of Asia hold in great reverence the children who have been nursed by their mothers and would not allow any others to inherit their father's wealth but those whom the mother had nursed. This is also why the Lacedaemonians, from the two sons Thomistes had left, elevated as their seventh king not the elder (because a foreign woman had nursed him) but the younger, nursed by his mother, the queen.[12] Their reason was very sound: it is necessary, for the son to worthily succeed the father, that he reflect his father's nature and virtue (without mentioning any suspicions that might arise when children are nursed by a foreigner, outside the house).[13] For it is easy to change a child when it is out being nursed. And indeed it is often suggested that those who do not reflect the behavior of their parents were changed while being nursed.

These are scarcely fitting inheritors of the wealth that in no way belongs to them. Consider the real children, tricked out of their inheritances, who then become artisans or workers (in whom one nevertheless notes a noble heart, a gentle and decent manner—for they naturally exude the goodness of their parents). This, in my opinion, is what has happened to most of those who seem both in nature and in behavior very different from their putative parents. It is from having been changed while out being nursed that this gentleman is completely surly, loutish, wretched, cowardly, and villainous, nowhere near the nature of those who think they have engendered him; and that this peasant is kind, decent, courteous, generous, and hardy, completely the opposite of those said to be his parents.

It is written of the good Artabanus,[14] the old and weary king of the Epirotes, that he had a son who was changed for that of a simple knight through the collusion of a nurse corrupted by money. Some time afterward, this nurse, having qualms of conscience, revealed the treachery, over which great wars were declared between different factions, both of whom in the end suffered great losses in a fierce battle. The kingdom was subsequently occupied by a foreigner (named Alexander, brother of the beautiful Olympias, mother of Alexander the Great). This desolation would not have happened if the queen, wife of Artabanus, had nursed her child.

This is why the most prudent legislators, Plato and Lycurgus, were justified in ordering all women of medium and low estate to nurse all their children, as much as they were able, and the great ladies and princesses to nurse at least their eldest.[15] It is a beautiful and a sacred law, which, if observed, would spare fathers and mothers much trouble and vexation over their children being ill nourished or being changed. For these things cause them so much pain that they would rather see them dead. What regrets decent, honorable, virtuous, modest, continent, and peaceful fathers and mothers have upon seeing one of their children insolent, drunk, gluttonous and lascivious, frequenting taverns, whoring, going to brothels, wandering the streets, gambling, cheating, thieving, swaggering, robbing, stealing, killing, quarreling seditiously, raving like an evil, perverse, and blasphemous madman given over to every wickedness! How heartbreaking it is for these good people to see themselves taunted and abused by this wicked wretch, if they are even able to stand having him in the house! Or if they turn him out, how heartbreaking to hear daily reports of his being sent to prison, to the galleys, to the gallows, or to the wheel! Of another child they might hear the disgrace of his having beaten or killed somebody and of his being sought everywhere, or that he has been charged with counterfeiting, buggering, or incest. Of yet another they will hear that he married a whore from a brothel, that he frequents the most vile wretches of the city, or that he takes part in every imaginable intemperance.

Although this happens every day, it in no way diminishes the extreme anguish these poor people suffer because they could not make their offspring virtuous, not even as children, on account of bad milk sucked from evil and wicked nurses in dissolute houses, and exposure to vile and indecent language and actions. Perhaps these children are not even their own, but other people's, ill-bred and ill-behaved like their ancestors. If they are incorrigible, it is in their nature or because of their first instruction, which leaves a very deep impression. If they are disobedient, it is because they have not had the model of a true father and mother raising them right from the beginning. They far more easily adopt the nature and behavior of their foster fathers and that of their foster mothers (most often wicked) than that of the men and women who know them to be their children.

I purposely remain silent about the other horrors that can befall the child, such as catching syphilis from the nurse. Here we see great disasters visiting much later an entire family when the father and mother, having on occasion put the child between them in bed, catch syphilis because it is still unapparent on the child's body. I say nothing of children unfortunately smothered by nurses who, very often overcome with wine, sleep too heavily. Such a mishap comes about rarely with mothers, because their natural love makes them more vigilant, diligent, and attentive in preventing such misfortunes. What a disaster it is, what regret, what upset, what rage, for a poor woman who has for so long wanted a child and done a thousand things so that she could have one, who has carried it within her womb with a thousand discomforts, who has brought it forth with the greatest pain and peril to her own life, when she has finally come through all these dangers, most joyous and happy to have a handsome son who makes her forget all the pain she had because of him, to be told only a few months later that his nurse smothered him!

I see, now, that all the women are converted and (thank God) most determined to nurse their children. There remains but one hindrance, and one that is not of their making. Women, realizing they are subject to their husbands (as they should be), ask to be excused from the conjugal bed because some husbands do not want to put up with the noise and racket that children often make. And so although they must keep separate rooms, the wives naturally do not leave their husbands of their own accord, for it is also ordered that no man shall separate those whom God has joined together. These women would be most willing to put up with the trouble children cause, so long as their husbands did not demand separate beds on these occasions.

There are also some husbands who do not wish to allow their wives to nurse so that their breasts will not sag but stay prettier, the way they want them to be for caressing. There are others who hate the smell of milk on their wives' bosoms. Rather fussy, are they not? And most of those who say such things make love to the nurse more often than to their wives. The sagging breasts of the nurse, or the smell of milk on her, do not disgust them. For such things they do not find her to be a bad wench. I dare speak of it further (think upon it, ladies): if your husbands do not want you to nurse, it is so they can have in the house another woman whom they think they can have under their command, so as to be unfaithful whenever they please. The reason they give for refusing to let their wives nurse is that they want to have a lot of children and that they would lose time in getting them pregnant again. Believe me, they also take a lot of pleasure in having a lot of nurses for the purpose of assuaging their carnal cupidity. For (as you know) nurses are easier to debauch than other wenches and servant girls. And one scarcely sees nurses going out of these fussy men's houses without their baskets full. And then they say it was the work of some valet or neighbor.

If women are circumspect, they will in decency keep their husbands from this mortal sin by allowing no nurses, either in their houses or anywhere else, and by performing this duty of nature themselves. And God will bless their efforts. As for husbands who cannot stand noise, sagging breasts, and the smell of milk, I will give them remedies for all these discomforts, if they would but ask.

CHAPTER II
WHEN A PARTURIENT WOMAN'S MILK IS GOOD; HOW MANY HOURS THE CHILD MUST GO BEFORE BEING FED; AND WHAT THE FIRST THING IS IT MUST BE GIVEN

When the child no longer has need of blood and is about to come out of the womb, the excess blood rises to the milk glands. The first of it to arrive there is disdained by the child as most foul and distasteful. This blood was furthermost from the womb and closest to the milk glands; upon rising, it is the first to flow into them. Such coarse and heavy blood makes itself into a first milk that is thick, murky, and full of curds, called colostrum by the Latins,[1] which from the oldest antiquity has always been considered foul and most pernicious, and so was forbidden to children during the first two days. It causes a stomach ailment called colostration, considered to be fatal. See what Pliny says about it.[2] Because of this, it is well advised that the parturient woman have a substitute woman (called a *soustenery* in Languedoc)[3] who gives her breast to the child during the first few days until this murky milk is drawn out by using a puppy, or some other means, and good milk comes into the milk glands, made from blood close to the womb, or at least better than the former, after the worst has been emptied out.

The Anabaptists do not allow any woman other than the mother to nurse the newborn, and so they make it suffer hunger for about twenty-four hours, saying that the child has no need of nourishment. For God makes nothing imperfect, and no necessary thing is ever omitted by Him. And so if the infant needed milk sooner, God would make it come sooner. It is, therefore, necessary that the infant scream and holler until the colostrum drains.[4]

Likewise,[5] the poor women of this region, and especially the village women, nurse their own children. They let their children suck the good and the bad, just as when they are bigger—even if the mother is pregnant, it makes no difference to them.[6] As long as there is milk, they give it to them, right down to the last drop, and they do not feel any the worse for it, since these children are of a robust constitution, born of mothers and fathers who were nourished coarsely. Such nursing cannot harm these people. But for city people, who are nourished more delicately, and for all who have the wherewithal to give their children better nourishment, this observation is most important and must be followed: for two days at the very least the child must not be nursed by its mother.

Must it be given the breast of its *soustenery* as soon as it is born? It is customary to wait a few hours before letting it nurse; some wait two, some three hours, some even more. For there are matrons who are of the opinion that the child must not be nursed until four hours after its birth. I would remind you that the fawns of animals, as soon as they are born, hasten to the mother's teats by a natural instinct and return to them hourly until their little stomachs enlarge and are capable of holding enough milk for longer periods. This is reasonable and in accordance with nature, since the child in the womb lives like a plant, constantly drawing sustenance from the earth through its roots. Thus, once out of the womb, it cannot last long without nourishment, so it screams and cries from hunger. This is why the fawn hastens to its mother's teats, with no fear of the colostrum (which animals also have); but they are less sensitive than our children. Also, since they do not have as much excrement, it is not so harmful for fawns to suckle immediately as it would be for our children who are born with their stomachs and their bowels full of a viscous and blackish humor, commonly called syroc, which must be expelled, at least from the stomach, before the child nurses. Otherwise, this humor would corrupt the milk that the child sucks. In order to make this humor descend and drain more quickly, soon after birth the child is given something appropriate, which we shall discuss presently.

Animals have none of these restrictions because they have no need of them, for (as we have said) they are not as full of excrement. Note that they do not have runny noses, spit, or weep, which are all means of expurgation. The substance of all this excrement goes into the making of fur, feathers, or scales. Man, who is born completely naked, is very soft and sensitive and does the most excreting of all the animals because he has the largest brain (being the most intelligent). Thus, it is important for the child not to suckle for two or three hours and, through a little crying, to exercise its lungs. When the lungs push down (through the diaphragm) on the stomach, they force out its excrement, warm the stomach, and make it ready to receive milk and make better use of it.

What is to be given to the child in the meantime to calm its insistent hunger, to which we alluded above? In ancient times they were given butter and honey, because of what was said to the prophet Isaiah, chapter seven: "Behold, a virgin shall conceive and bear a son, and his name shall be called Immanuel. He shall eat butter and honey . . . "[7] I understand that even today the Jews give their children a little of each before they begin suckling. As for our own, they are given various things: some give theriaca, mithridate, or the stalk of a bean; others give a spoonful of rose honey, violet syrup, or a little powdered sugar with very thin gold leaf cut into small pieces, and others yet give something else, such as, in the region of Agen, almond oil with cane sugar, and the same is given to the mother; or a spoonful of pure wine, as do the common people in France; or crushed garlic, to get them accustomed to it early, making them less subject to vermin.

Those who give them theriaca or mithridate think that the syroc, which children have in their bodies, is a poisonous substance because it is blackish and ugly looking. But it is merely an excrement resembling the foul matter of the bowels that will be forming there some time later. This is why rose honey and violet syrup are very good and are appropriate for purging and ridding the child of this matter. In order to treat both conditions, I usually prescribe sugar and gold. For the sugar purges and cleans a child out well enough, while the gold counteracts the poison. Besides, this treatment is in accordance with popular opinion.[8] Thus, shortly after the child cries, he is to be given one of these remedies, and two hours after that it can be nursed, even after having slept.[9] But it must abstain from the mother's milk for the first three days.

CHAPTER III
THAT A VIRGIN IS ABLE TO PRODUCE MILK IN CONSIDERABLE QUANTITY

Logicians use faulty reasoning when they say, "If she has milk she has had a child," since pregnant women, before their delivery, can produce fair amounts of it. They reason much more properly when they infer from the milk that she has known a man. Yet such a rule is not so hard and fast that on occasion things do not happen otherwise. For if one squeezes the milk glands of newborn children, a little milk will come out in most children, if not in all.

But I am not stopping here. I wish to prove that, in older girls, over twelve years of age, large amounts of milk can be found, even if they are virgins. Hippocrates was the first to make us aware of this, writing in his *Aphorisms* that if a female, without being pregnant or having delivered, has milk, her natural purgations are being blocked.[1] The reason is most evident for those who know where milk comes from, and when we will have revealed it, this statement will not be as strange or unusual as it first seems.

We pointed out in the first chapter of the second book that the female sex, comparatively cold and moist,[2] has more blood than the male, but it is less refined and watery. Nature made it that way in order to provide nourishment for children, which women must normally carry for nine months. Children refine it further in their livers, which must not be left idle. The mother would be unable to generate the quantity of blood required if it did not remain imperfect. The father has less blood, but it is richer and more refined because of the semen that must come from it. He must furnish a more efficacious blood than the female. Women, therefore, have a lot of blood, since it sometimes supplies two, three, even four, and on up to nine people, according to the number of children in a burden.[3] And in women who are not pregnant but very healthy, a portion of it remains superfluous and becomes excrement, merely by virtue of its quantity.

This excess can only be harmful to the body, bursting the veins or smothering the natural heat. So nature compensates for it by allowing the heavy and less-refined blood to be separated and expelled through the veins of the womb once each month, following the cycle of the moon. This has given rise to the common saying that women take after the moon and are governed by it, as Aristotle also has said.[4] What they void is completely useless because they have a much greater supply of it than their bodies need before conception. After conception, everything is usually retained in order to nourish the little one, who takes full advantage of what was excessive to the mother, putting to his own use the pituitous blood, refining it for himself. When the child is larger and gets ready to come into the world, nature, which took care to furnish the child's dwelling place with food before its birth, suddenly thinks about nourishing it in its first years with a substance both suitable to its delicateness and akin to the food it took in the womb. For its fragility could not stand an abrupt change, and it needs a very pleasant food, especially since it must be taken in through the mouth and not the navel.

For these two reasons it has been ordained that the blood that is left over, serving no purpose in the mother, will rise up to the milk glands after childbirth instead of being expelled each month. There it becomes sweeter and whiter, fashioned by glands that nature has put there in great number for that very purpose. These glands treat with their heat (and transform according to their nature) the phlegmatic and imperfect blood brought to them from throughout the body. One must not believe what our predecessors did: that there were certain vessels that, through a continuous passageway, carry directly to the milk glands the blood that previously flowed into the womb, and by means of which it founded an accord between these organs.[5] It is true that the flux down below usually stops while the woman produces milk, but the passage from one place to the other continues through long wanderings in the huge vena cava and in its ramifications until the blood finally arrives in the branchings that bring the nourishment to the chest and to the breasts. Those who believe that milk is made from blood that is decomposed upon entering the milk glands err gravely also, for it was already only half-processed, highly moistened with an insipid natural pituitous quality.[6] The milk glands work upon this blood until it becomes thick, sweet, and perfectly white. These qualities come only from concoction, which ordinarily ends up in assimilation, the ultimate objective of nature. But such ideas are better kept for promulgation in our schools (where it is necessary to point out the errors of ordinary physicians) than given as instruction for laymen. Let us return to our discussion and conclude forthwith and without further argument what we have proposed.

As females spend energy growing, their veins contain much more blood than is needed for the sustenance of their bodies. This is why it accumulates around the womb, and why that which is excessive is expelled from the womb at certain intervals. If a woman happens to conceive, everything is retained by the child. After birth, it is retained to make milk. If she does not conceive and yet does not continue to have her purgations each month (as she usually did), we think that her blood is diminished for some other reason, and that there is nothing left over from it after her body has taken as much as it needs. It could also be that the veins of the womb are closed and stopped up with some thick matter that stops the blood from flowing out, or that the blood is misdirected elsewhere and causes considerable damage there. Sometimes women have ugly red spots on the face because of blood that frequently rises to the higher parts. Some have headaches or feel bloated because of its quantity or vapors.[7] Some faint, some become crazed; some have nosebleeds, while others vomit blood. Some have trouble breathing because of pulmonary repletion, and some have backaches because of excessive blood in the great vein. Others cannot walk because of a heaviness in the legs due to no other reason than an excess of blood.

And so, the bosom, receiving great amounts of blood in a small amount of time, can grow in size and swell out of proportion. Likewise, one notes that as soon as the body stops growing and begins to abound in blood, the bosom becomes round and full as the milk glands push forward and swell. If, therefore, they then receive more blood than they require for their nourishment, they will increase in size in every visible dimension; and if the cause persists, why could the milk glands not make milk from what abounds in them, since they are endowed by nature with such a property?

Someone might reply that the milk glands do not undertake such action except to nourish the child born from the body to which they belong. But this would be suggesting that our organs have some discretion or power of reasoning, which is false. One would far better argue that, in spite of the affluence of blood, the breasts will not make milk from it if they have not received through conception a certain quality exciting their lactiferous capacity. But this second reason, founded solely on the experience of what happens most often, cannot overturn the first. For if the glands of the breast, because of their constitution and structure, have the capacity to convert blood into milk, provided they are supplied with more than they can consume (this is why we say that milk is their benign excrement, as the substance of semen is that of members), why would they not do so every time this happened? If this ability came from the child, it would not be inborn, as we consider it to be. Furthermore, when wet nurses lose their milk we can make it start up again long after their deliveries by bringing blood to the breasts. And stop to think! Aristotle clearly says (and it is, in fact, also seen to be the case) that some men have milk, which can be sucked or squeezed out.[8] The story is also told of a Syrian man who nursed his child for more than six months with his own milk. There is, therefore, nothing to stop a female from having milk, without ever having given birth or conceived, through the sole retention of her menses. All that is required is that the frenzy of the blood push into the milk glands.

But, in truth, this condition lasts but a short while, only during the few surges of blood that storm the milk glands. Soon afterward it leaves for other organs, if it is not maintained therein by frequent attraction, or if it finds an exit through the veins of the womb. This is why it is a rare thing to see a young woman producing milk. Nevertheless, it can happen for the reasons mentioned above, which confirm Hippocrates as correct in the aphorism we cited. One must not, therefore, deny virginity, without due consideration, in a woman who has produced some milk, since the authority of such a great person (who may well have foreseen just such a case) is able to cast doubt on our judgment. The jurisconsult rules accordingly, basing himself solely upon the authority of Hippocrates, in part seven of the seventeenth book of *Paul's Responses* in law *septimo. ff. de statu hom.*[9] But the fact of abundance is stronger than all the authority of the most learned people in the world, and it seems to me that the arguments given demonstrate fairly clearly that, without having been pregnant or having given birth, but only because of the repletion of the veins in the breasts (a result of the suppression of the flowers), it is natural for the female to produce milk, which will, if sucked, continue for some time.

Some time after this discussion, I found myself in Gardonne in the good company of some gentlemen at the house of Monsieur de Longa, governor of Bergerac, among whom was Monsieur Menier, a learned and expert physician, who affirmed along with several others there present that in a village near the town of Saint Cybran (that is to say, Cyprien, in proper speech) in the region of Périgord, the following case arose. A poor woman died, leaving her five- or six-month-old child without a nurse. Her niece, verily a young girl and a virgin, a plump child and in good health, tried a few times to calm the child (which was dying of hunger) by offering it her breasts as if there were milk in them. The child so pulled and drew on them that it made some bloodied humor come out, which the poor girl endured patiently in order to please her little cousin on whom she had pity. This went on so much that the red turned to white and became milk, with which she eventually nursed the child for more than a year, without fearing reproach for any sort of dishonor on account of the good witnesses she had to the source and circumstances of this nursing, which was attributed to the extraordinary providence of God and to a true miracle.[10]

CHAPTER IV
WHETHER THERE IS CERTAIN KNOWLEDGE OF THE VIRGINITY OF A MAIDEN

This subject is not in its proper place here, in a discussion of milk and the nourishment of children, but since we have just raised the question of a virgin's ability to produce milk, as well as the notion that one cannot argue for a maiden's corruption because of the presence of milk (contrary to popular opinion), I thought it logical to consider now whether there is any certain argument for virginity.

In my opinion, maidens should abstain from reading this chapter,[1] as should women who are squeamish about watching the anatomy of the shameful parts of their sex, even though one can read these words with less embarrassment than watching publicly what they stand for.[2] Yet, does one not often see some of the most decent, chaste, and modest women attending public anatomies of the shameful parts, and hearing them spoken of distinctly and without the slightest taint to their decency? One ought not spare the ears more than the eyes. At least those who do not fear seeing in public those shameful parts will not fear hearing them spoken of, and still less reading this chapter in private. Stop to think! There are many proper and decent women who, in order for some ailment or condition of their womb to be more precisely recognized, will indeed allow surgeons (for a reason that convinces them to do so) to uncover their shameful parts and, in the presence of a physician, examine them closely with a speculum, which permits us to see right to the bottom of the passageway. Why, then, would they fear hearing about this mystery of virginity, or how will they be able to find evil my treating it in great detail? Still, I am not upset about all this, since these words are only for the judges, and especially the midwives.[3] As for the others, the subject will remain a secret if it is disgusting to them.

But the question is of great importance, for the honor and dishonor of maidens, for the dissolution of contracted marriages with an impotent, cold, or deformed man, and for the condemnation or acquittal of the man accused of having overpowered, raped, or deflowered a maiden who was not willing. This is why magistrates, and even more the physicians and surgeons of these deputies in whom the magistrates put their trust, must take great care. And so if there is an error made, the fault is to be blamed more on experts who gave false information than on the judge who did the sentencing. The matrons or midwives attribute to themselves the prerogative of knowing better than we or than surgeons how to pass judgment on a maiden's virginity, especially since they are more experienced and accustomed than men, having familiarity with maidens and free access to them—those who are untouched as well as those who are corrupted, both of whom communicate with midwives more willingly than with men (even though men are more knowledgeable).

But matrons can be gravely mistaken in these matters, especially since they are not well versed in the anatomy of the shameful parts. For only he who has had a great deal of practice in the ocular examination of wombs of all age groups can know the truth of virginity. Hippocrates says of medicine in general that judgment is most difficult.[4] I say, likewise, that it is most troublesome to decide on virginity, and still more to vouch for it, following what is written about it in Aesop: a man who had always carried his twin daughters in a bag about his shoulders from the time they were born, upon being asked if they were virgins, answered that he could be sure about the one he carried in the front but not about the one he carried on his back. It is a hard animal to keep watch over, as the proverb goes. And as for the certainty, either of deflowering or of virginity, midwives are sometimes too quick to offer it. I find much greater difficulty in it, even though I am not ignorant of uterine anatomy, as they indeed are for the most part. But I do wish to except at least Mistress Jane Massale, known as Gervaise, matron of Montpellier, truly a learned woman, and one who hardly ever fails to attend public anatomies when we have a female to dissect.[5]

Now, in order to show the abuses committed during inquiries into virginity, I shall divide the signs and lines of reasoning given by laymen into two types: first, those that are among the most foolish, sought in the face, on the neck, on the breasts, and elsewhere, without examining the shameful parts; and then those sought more properly in the depths of the shameful parts. In discussing the latter, I shall cite three depositions sworn by matrons[6] in order to show their agreement on the main points they touch upon.

One of the signs people believe to be among the most telling is absurd in the extreme.[7] It is that the nipple, or small end of the teat, changes color the instant a maiden is deflowered. Its surrounding area supposedly becomes brown or blackish or is otherwise changed. But there are many old women who are indeed virgins and yet have them thus colored! This is common to all females, and it is merely through aging that this surrounding area (called by the Greeks *phos*, which also means "light") changes color. And how would it be possible for this change to take place instantaneously upon the opening of the chamber of maidenhood? What would be the immediate, remote, ultimate, and adjacent cause?[8] I will grant that there is a very close accord between the breasts and the womb, as I have pointed out in the preceding chapter and shall be able to do even better in the next. But the closest accord there could possibly be between all the organs of our bodies could not cause such a change, nor so suddenly, even in the matter of color. Deflowering would sooner be recognized in the face and in the eyes, if the maiden is not overly bold, brazen-faced, and shameless. For upon being deflowered, even though this be done decently and in marriage, she is a little subdued and ashamed; her eyes are sad, dulled, and demure; her face blushes easily when she sees those who are dear to her. These are changes that can come about suddenly in maidens if they are modest and decent. For just the day before they were seen to be more cheerful and playful; yet as soon as they lose their maidenhead they are of another countenance, and their faces are completely changed. But as for their breasts, it is pure foolishness what is said about them.

Equally inane is another sign people believe common in boys and girls who have lost their virginity. Measure with a string around the neck, then from the chin to the top of the head. If the dimensions are the same the person is a virgin; if the neck is bigger the person is corrupted. For (so they say) the neck thickens the instant one is corrupted, whether by one's own hand or by another person. But this cannot take place at the defloration of a maiden any more than it can happen at that of a boy. Nor is he considered any less a virgin because of the nocturnal emissions he might have had. Moreover, there is nothing to discuss, since in puberty the neck thickens by itself; and it is at that time that the child's voice changes (which in Greek is called *tragan*, meaning *bouquiner*),[9] because the trachea or throat visibly thickens on account of the stronger and drier heat. It thus follows that the neck gets larger also. And who would doubt that some of them remain virgins long after the end of puberty?

People also say that the instant boys or girls lose their virginity, the ends of their noses open slightly, and that from that point on a manifest separation is found between the two pieces of cartilage. But this is a fabrication, for the division is always there, and one feels it more sharply when the body is dried out. It happens in puberty and afterward that hair also appears on the shameful parts, giving evidence of the considerable exsiccation taking place. Thus, those men who sooner give themselves to women will sooner have a beard than they would otherwise, since their body dries out more. Martial speaks thus on the subject:

> Whence the goatlike one, and the wild hairs.
> The mother is overjoyed on seeing a beard on her son.[10]

Tests are also given to find out if the maiden is a virgin. Give her a little powdered lignum aloe to drink or to eat; if she is a virgin she will piss immediately. Item: put on hot coals some broken patience dock leaves,[11] and have the maiden smell the smoke; if she does not bepiss herself she is not a virgin; likewise if she does not become pale from patience dock flowers. All these tests are ill founded and one ought not to pay them any attention whatsoever.

It is now necessary to come closer and to descend into the infernal abyss of the most devout Alibech of Boccaccio, into which the good and saintly hermit Rustico put his devil.[12] It is there that the secret of maidenhood will be found, and where one will learn about it. This was something very well understood by the prior who (as it is written in the *Heptameron* of tales of the queen of Navarre)[13] called himself a visitor of bodies as well as souls, claiming to a young nun that her own abbesses and prioresses had passed through his hands in order to have their virginity visited. Not stopping at the signs one might read in the breasts (where he nevertheless wished to place his hands) or on the neck, or in the nose, he wished rather to see around where she pissed, ordering her to lie down on a bed and to cover her face with the front of her habit because (he said) the thing remained to be proved.

We will now consider the second type of signs and lines of reasoning proposed in order to recognize defloration and virginity. And first let us hear what midwives say about it. I have two sworn depositions, one from Paris, the other from Béarn, which are far enough apart so as not to be in communication. From these depositions we will be able to see to what extent these women are in agreement in their signs and in their thinking (which ought to be uniform if they are truthful—for truth is consonant and in agreement with itself).[14] And each woman's erotic parts are similar to every other's, whether she be from Paris or from Béarn, or any other place in the world; and whether she be a gentlewoman or a peasant, beautiful or ugly. For (as vulgar people say), cover up the face, the rest is the same. It is but the complexion and the features of the face that amuse and abuse men (unless perhaps it be the grace, countenance, and speech that draw us more to one and cause her to be loved more than a prettier woman without these other pleasant qualities).

Let us see, then, how these reports concur, one concerning defloration, and the other, virginity. For they must be harmonious by virtue of the law of contraries.[15] And let us begin with the one from Béarn, because it attests to virginity, which is the first in time, in order, and in dignity:

We, Jouanne del Mon, and Jouanne Verguiere, and Beatrix Laurede, of the parish of Espero in Béarn, matrons and sacristans, interrogators and examiners, certify to all concerned that by a decree of justice and by order of the high magistrate, His Honor the judge of the aforesaid place of Espero, on the fifteenth day of the month of May in the year fifteen hundred forty five, we the aforesaid matrons have found, examined, and scrutinized Marietta de Garigues, fifteen years of age or thereabouts, because the aforesaid Marietta maintained that she had been assaulted, forced, and deflowered. Whence we the aforesaid sacristans examined and scrutinized all with three lit candles, touched with our hands, saw with our eyes, and probed with our fingers.[16] And we found that the *podads* were not rent, nor was the *halhon* displaced, nor the *barbolo* beaten down, nor the *entrepé* wrinkled, nor the *reffiron* opened, nor the *gingibert* split, nor the *pepillou* flattened, nor the *dame dau miech* withdrawn, nor the *tres* out of place, nor the *vilipendis* scraped, nor the *guillenar* dilated, nor the *barrevidau* turned up, nor the *os Bertrand* broken, nor the *bipendix* chafed. The whole we the aforesaid matrons and sacristans thus state by our report and fit judgment.

There we have fourteen points representing virginity, according to the women of Béarn. Let us now see the deposition of the Parisian women, who give their report of a maiden who was deflowered:

We, Marion Teste, Jane de Meaus, Jane de la Guigans, and Magdeleine de la Lippuë, sworn matrons of the city of Paris, certify to all concerned that, on the fourteenth day of June 1532, by order of Monsieur the Prevost of Paris, or his lieutenant in the aforesaid city, we transported ourselves to the rue de Frepaut where there hangs a sign of a slipper, and where we saw and examined Henriette Peliciere, a young maiden fifteen years of age or thereabouts, on the grounds of a complaint lodged by her in justice against Simon le Bragard, whom she accused of having forced and deflowered her. And all having been seen and examined with the fingers and with the eyes, we find that she has the *barres* broken,[17] the *halerons* displaced, the *dame du milieu* withdrawn, the *pouuant* rent, the *toutons* out of place, the *enchenart* turned up, the *babolle* beaten down, the *entrepet* wrinkled, the *arrierefosse* opened, the *guilboquet* split, the *lippion* flattened, the *guilheuart* dilated, the *balunaus* hanging down.

There we have fifteen, counting correctly, which correspond rather well to the fourteen signs of the women of Béarn, as I compare them to each other, except for the last one, *balunaus*, which does not have its counterpart, as far as I can tell:

1. *Brocadés podads.*	*Pouuant debiffé.*
2. *Halhon delougat.*	*Haleron demis.*
3. *Barbole abaissade.*	*Barbolle abbattue.*
4. *L'entrepé riddat.*	*Entrepet riddé.*
5. *Reffiron ubert.*	*Arrierefosse ouverte.*
6. *Gingibert fendut.*	*Guilboquet fandu.*

7.	*Pepilhon recoquilhat.*	*Lippion recoquilhé.*
8.	*Dame du miech retirade.*	*Dame du milieu retirée.*
9.	*Tres desutades.*	*Toutons deuoyés.*
10.	*Vilipendis pelat.*	*Lipandis pelé.*
11.	*Guileuard alargat.*	*Guilheuart elargi.*
12.	*Barreuidau desutade.*	*Enchenart retourné.*
13.	*L'os Bertrand romput.*	*Barres froissées.*
14.	*Bipendix escorgeat.*	*Barbidant ecorché.*

I wish to add a third, which is the deposition of the matrons of Carcassonne, for even greater confirmation of these terms. For it is said that in the mouth of two or three resides all truth:

> We, Guilhaumino and Iano, deputies of the low part of the city of Carcassonne, officially present by request of the officer of the aforesaid Carcassonne, in order to examine Marguerite d'Astorguin to determine whether she was deflowered, state and swear to all who will see and read these words that on this day we transported ourselves to the house of the aforesaid D'Astorguin and there found her lying on a bed. After having three wax candles lit, we scrutinized her with our eyes and touched and palpated her with our fingers. Having found that the *os Bertrand* is broken and split, the *danno del miech* is withdrawn, the *tres* out of place, the *quinqueirol* shifted, the *intrans* and the *pindourlets* all chafed, the edges of the *coustats* worn down, the *pels* on top all flattened. Because of this, we state that the aforesaid Marguerite, on account of a man's prick having passed in these parts, is indeed deflowered and devirginated. And such we state and attest.

Now, let us begin examining these arguments or signs. Some are very flimsy, and others are false. Flimsy are those that refer only to some pressure brought to bear upon the particular shameful part. A most false sign is that of the broken *os Bertrand*,[18] for we demonstrated in the first chapter of the fourth book that even in childbirth (which is a much greater strain) it does not open or break. Let us leave undiscussed the other signs put forth by the matrons, in terms that are proper and peculiar to them alone, which are like terms of the trade, or a jargon of their craft, understood by very few people, and let us come to the main point, which from time immemorial has been renowned as the true mark of virginity.

It is the *dame du milieu*, which the ancients called the hymen, belt, or zone, and cloister of maidenhood. It is a membrane stretched across the passageway, and which must be broken in defloration. And for this reason the god who presides over weddings is called Hymenaeus, and he is invoked to be propitious to virgins in this combat so that they will not die from it. Many consider it to be a poetic fiction and an error on the part of people poorly versed in anatomy (be they physicians or surgeons) who have learned and held, up to the present, that there is, in front of the neck of the womb, about in the middle of the passageway devoted to the virile member, a membrane composed of veins and arteries in the guise of a hedge which is broken during defloration. As a result, the poor girls have great pain and lose a little brilliant red blood.

The modern physicians, Fernel, Sylvius, Vassaeus,[19] and others, consider this to be a fable, affirming that there is no obstacle, diaphragm, hedge, or little wall (however one might want to call it) in this passageway, not any more than there is in the rectum (too well known by abominable sodomites). If such were true, the pain that a virgin feels during her defloration would only be a result of the extension or dilatation of the conduit (which, up until then, had remained constrained and tight, and which is now enlarged through force, such as when one puts one's finger in the fundament of a small child in order to probe for a stone).

Since a maiden's passageway is dilatable almost without limit, one must not find it unusual to hear that some girls have been deflowered at six or seven years of age (and even younger still) by infamous villains. Of course, the narrower the girl is, the more pain she will suffer upon the first entry of the member, which forces her to become wider. A similar pain (but a little sharper) is felt in childbirth, for which this passageway must be still more dilated. But then everything goes back into place and tightens up nicely after the child has come out, so much so that the conduit is, as a result, hardly wider than it was before. It is like a muscular and thick bowel that is able to widen under pressure, and when the pressure is released, return to its former size, or very close to it. It is quite true that the woman who has never carried children (even though her instrument has been frequently and for a long time visited and reconnoitered)[20] remains tighter than if she had had children. But some women, after having had several children, will not be wider than others who are recently wed. This is as much a result of the woman's size and her structure[21] as it is of her corpulence, in conjunction with the caliber of the member enjoying her.[22]

For, regarding size, is it not reasonable that a larger body would have all of its parts larger, too (if it is well proportioned), and consequently more roomy natural openings? And in bodies that are less well proportioned, does one not see on some a very large mouth, split all the way to the ears; and on others, large and wide ears, like fans for winnowing grain? There are some who have huge, wide-open eyes; others have nostrils so open and patent that one can see all the way to the brain, so to speak. Some have very long fingers, very long legs, and short bodies. Others, on the contrary, have everything small and with little openings.

Likewise with internal organs. Some have a huge and capacious stomach, able to hold a great amount of food even though the body might be small; others have a large liver. There are some who have very capacious bladders, huge bowels, wide veins and arteries; others, on the contrary, have everything more reduced, or this organ narrower and that one less so. Why would it not be the same with respect both to the womb and its passageway, as is also the case with the virile member, which corresponds to it proportionately? Do all men have the same size or caliber, in every dimension? It is certain that they do not, even though one may say: *Ad formam nasi cognoscitur ad te leuaui.*[23] And this is because the proportion of members is not observed in all men. Several who have a stunning trunk of a nose are flattened elsewhere, and several who have flat noses are well endowed in the principal member.

It is said that women who have a large mouth are also larger elsewhere, and those who have small feet are narrow in their privities.[24] Perhaps that would be so if everything were proportioned the same, which is not always the way things are, as I have demonstrated. This is why one often finds it to be just the opposite of what is commonly said. It very frequently happens that, according to their size, large women have everything larger, and small ones, everything smaller, and that as to the structure of the parts (respecting certain proportions throughout the body), from the large opening and capacity of one of them, such features of the others are understood and inferred, but not always and in all women.

And for this reason, we add to it the notion of corpulence, which is of great importance in this matter. For women who have firm flesh have tighter privities, and flabby ones, the opposite. Finally, a member of higher caliber makes for a greater opening and dilatation than a small one, because this sheath only widens to the degree of the instrument it houses. Since, therefore, the size and corpulence vary, since the structure of the organs varies as a result of different degrees of fleshiness, and since maidens of the same age differ in the capacity of their wombs, how is it that, when Rustico's devil has been through them and they still remain different according to the caliber of its hornless head, one will be able to determine their virginity by probing them with the finger, or with a candle, or with a speculum, in order to see if the passageway is more or less tight and narrow or loose and wide? For if the maiden is of a nubile age and of the corpulence required for marriage, she will accommodate without difficulty, even though she still is a virgin, a fairly large probe, just as she would well accommodate a man's member[25] that is equal in size. Yet one will not say that because of the candle the maiden is any less a virgin, as one would certainly maintain if the member[26] had been inserted.

And what difference will there be in this conduit?[27] Will it not be the same as before, with the same shape, disposition of parts, and other accidents, for having accommodated the candle as for accommodating the virile member, and vice versa? This is how certain maidens are wronged, by thus probing them in order to see whether they are whole or uncorrupted; for if the candle goes in rather easily, it is judged that the virile member has already entered, and yet there will be nothing established other than the fact that her conduit is easily dilatable, and that the candle could just as much have been the first as could the suspected member. And if, upon probing her, the conduit is found to be very narrow, such that the candle will go in only with great difficulty, what will be said? That she is a virgin? Certainly, but she will not be one any longer, after the candle has been inserted. For probe her once again and the probe would go in so readily that you would judge, on the contrary, that she is not a virgin. Likewise, if for some reason one had to insert pessaries because of retention of overdue flowers in an older woman, or because of some other indisposition common to virgins, you would still find her a virgin. But how would you be able to tell that the opening was attributable to the virile member rather than to a candle, or a pessary, or the finger of the maiden herself, foolish enough to have corrupted her own body?

No trace remains to point out these differences. All these maidens, then, will be equally deflowered. And there will be others who will not be considered virgins, even though nothing has ever been inserted into them, simply because with the first test the tube is found easy to dilate and quick to yield because of its capaciousness and natural suppleness, as in those who are big-limbed and especially well reinforced. And some other wretched maiden, very tight by nature but in fact deflowered, will be considered a virgin by the aforementioned test, as well as by the following one, which must not be neglected.[28]

Another sign among laymen, usually used to determine virginity, is based upon the maiden's manner of pissing. The virgin's pissing is more unfettered and clear than other women, because her womb pipe is still tight and narrow, all the way to the outside end, which makes her piss straight and far, in rather the same manner as a man, whose urinary canal is very narrow. Thus, once her privities are widened, by whatever thing it might be, she will piss like a corrupted woman and will have lost this beautiful mark of maidenhood, all the while remaining a virgin, that is to say, not known by a man, to whom one will want to attribute her defloration and, consequently, force him to marry her or to dower her in accordance with her station.

On the other hand, a small maiden of fourteen or fifteen, deflowered by a man with a small member, will appear more a virgin by every test than another of a larger size, twenty-five years old, who is truly a virgin and who was tested. For the heavy corpulence and the well-furnished buttocks and hips benefit the womb, very roomily housed and easily able to dilate. One must not be content, then, with this sign of narrowness, which varies greatly in various maidens, and in women also who have used the male for a long time, and even (moreover) in women who have had many children. The reasons for this are so evident that it is not necessary to discuss them in further detail.

Let us come back to the *dame du milieu*, which is like a loophole in the dike and which must be broken by the first to make love's assault and combat. We have said that several deny the existence of this enclosure or defense, and I was of their opinion for a long time; but in the end I was made aware of it by Fallopio,[29] and I looked into the matter more closely and explored more seriously what he writes about it in his careful *Anatomical Observations*. I find that behind the conduit of the bladder, through which the urine empties into the large canal, there is on each side of it a fleshy membrane in the shape of a half-circle, and that both of them join to close the large canal. Their conjunction is aided by a certain viscosity resembling the gowl that collects and sticks the eyelids together. It is not a continuous membrane as several have believed, but two contiguous ones joined with some sort of glue, and by means of which the large canal is slightly closed so that when the menses happen along, a small passageway is made in the very center, through which the menstrual blood beads and drips out.

But when a maiden is deflowered, the virile member makes a complete opening by pushing these two membranes all the way back against the sides of the large canal, where afterward they remain withdrawn and flattened without ever returning and sticking together again. And this is what matrons call the *dame du milieu* withdrawn. The vestiges of it can still be seen in old women even though they have had many children, but it is nothing but a small thread on each side, the rest having been lost and (as one might say) worn away for having been rubbed and rubbed an infinite number of times.

Soranus[30] is of a different opinion, claiming that the hymen is nothing more than a narrowing in this canal, made up of certain folds, formed and gathered together by a few vessels, which come there from the womb. So this canal is like a purse whose pleats are gathered together with threads or strings, which break upon opening if they do not give way. Thus, Soranus thinks that there will be pain upon deflowering because these folds are stretched by the rupture of its vessels, from which some bleeding occurs.[31]

But according to us,[32] the pain a maiden feels upon being deflowered is because the man's stick[33] does not separate these membranes little by little but forces them abruptly with its head, which is bigger around than the rest of it. For those husbands who think they will never get it in too soon (and still more those lascivious violators of sacred maidenheads) go about it like brutes, wanting to enter all of a sudden. If one but tried to separate little by little these two membranes [first with a small member, then with a medium one, and finally with a bigger one (if one had three of them, as the fellow feigned whose spouse so feared the big handle and then found it was too little)],[34] the maiden would certainly not suffer pain, just as without any pain one separates little by little eyelids that are gummed together. If glued eyelids are opened all of a sudden, besides causing a lot of pain, either one or both of them could tear because of the viscosity keeping them firmly stuck together. Gummed eyes must first be soaked and then each eyelid gently pulled up or down.

This is why several maidens suffer hurt and dilaceration upon the opening of this passageway, and why one of the membranes rips away a portion of the other. Such things happen more often to those who are old than to young girls, especially since the glue becomes stronger as the body dries out[35] and, consequently, holds more strongly, so that if one goes about it abruptly, one damages everything. Because these membranes are stronger and harder, and because their glue is drier, the womb pipe and its adjacent parts sometimes swell up so much from the violent impetuousness with which some proceed that for a long time afterward the member cannot be inserted. In young maidens who are still soft (called delicate ones), the membranes are tender and their glue is only mucous and runny, and so if one goes about it carefully there is not much difficulty, provided still that the subject be of the required size and that one has but to separate and push back these membranes, which are actually valves, that is to say, gates split in two and that fold inward. Hence, one can call the canal leading to the womb and providing access to it the vulva. The womb is like a small room prepared for the harboring of the child, and it has an antechamber between it and the large canal. The vulva is the true neck of the womb, of which we shall presently be speaking.

Now, from all this one can see how and why several maidens lose some blood during their defloration: namely, because of the dilaceration of this hymen, especially in those who are older. Younger ones can bleed also, even if they have already had their menses a few times. For behind these small membranes is retained some residue of blood that has flowed down from higher parts. And when a large opening is made, this residue comes out with the first assault through the new breach. This is how all maidens are able to have some bleeding during their defloration, provided they are in their puberty and able to menstruate. So, it is more reasonable not to marry off one's daughters any sooner, according to nature's laws written in our hearts. And I think that the law of God does not permit it any other way. Thus, not without reason is it said in Deuteronomy[36] that if a woman is accused by her husband of not being a virgin, her mother and father will present to the elders of the city the bedclothes or sheets, on which there will be found signs of her virginity. From this it can be seen that the parents were careful to keep the linen and the nightgown of the wedding night to prove and fittingly answer for the virginity of their daughters. Even today the Spaniards, very observant of such ceremony, have the matrons show, the morning after the wedding, in public and with great acclamation, the sheets of the nuptial bed so as to make apparent the stains of the defloration, crying out several times from a window that opens on the street: "*Viergen la tenemos.*"[37]

But there can be a lot of deception, for (according to the proverb) one is more deceived in women and horses than in any other animals. This is sufficiently proved to be the case by the fact that the spirit of God has provided for proof of virginity in Holy Scripture. And so, I leave aside the authority of so many knowledgeable physicians, both ancient and modern, whom I could cite and who are of the same opinion. For the word of God, which has created and formed all things, is beyond comparison better able to convince and assure us.

Yet for those who are more curious, it will be permitted to see what is said on the matter by Avicenna (in his third book, leaf twenty-one of the first chapter of the first part), by Alessandro Benedetti in his *Anatomy*, and also by Carpi,[38] who recites this verse:

> It is most wrong, and the meanest mistake,
> If a maiden's hymen you should break.

Supplement these lines with "outside of marriage." Also, Vesalius, in the second edition of his *Anatomy*, recognizes the hymen, and since then in his book *Radicis chynae usus*. Celsus, earlier than any of the above, seems to have understood as much when he goes after kidney stones in women from the front, but in virgins from behind, just as in men. Either he thought they could not be probed from the front because of the hymen, or he did not want this beautiful mark of maidenhood to be corrupted in virgins.[39] This is the more likely, for the finger of the surgeon could enter there just as well as the virile member. In this manner he discreetly reproached those foolish maidens who do not have such concern for their honor.[40]

There is another cloister or enclosure (called *reffiron* and *arrierefosse*[41] by the matrons) that is of no less importance than the hymen, and even more so in my opinion. For the membranes and valves mentioned above can be opened and pushed back by the maiden herself upon inserting her finger often. This is the case with a few who are unchaste of heart, and who would willingly receive into their hell the good hermit's devil if they but had the opportunity and were not kept from it through fear and obedience. Such maidens are off to a bad start with an evil inclination to lasciviousness, either because of idleness, foolish company, books of love, or yet other sources of lasciviousness.[42]

But there is another stronghold and ravelin deeper within which the maiden is not able to reach with her fingers, or at least not able to open without some other means. It is the antechamber we spoke of above, properly called the neck of the womb,[43] which is split crossways, contrariwise to the hymen and to the shameful part first seen from the outside. For there are three gates to the womb: two in the shape of valves, and the third, split crossways. This neck of the womb is round and hard, resembling the head of a lamprey, thus split and rough, as if it were garnished on the inside. It is necessary for this conduit to open for conception. Indeed, the sperm can be cast into the large canal as much as one wants, but unless it enters into this opening nothing will happen. This passageway is the tightest and opens the last of all. Sometimes one will have enjoyed a maiden for a very long time before the neck of the womb will open. Hence, since carnal copulation has generation as its end and principal objective, she can still be called a virgin because of this second maidenhead.

Besides, the greatest pleasure to be had in the venereal act is in this particular place. This is why all the rest can be nothing but foolishness, and not in good faith. This (in my opinion) is the main cloister, or *arrierefosse* of virginity, and a maiden must not be considered deflowered or devirginated as long as this *arrierefosse* has not been opened. It is like the outer wall one encounters after having leapt over the last ditch: one still has to break through it if one wants to enter the stronghold and plant one's banner.

Now, one can tell whether this *reffiron* or *arrierefosse* (as the matrons call it) has at some time been opened, by one of two means. One is by dilating and widening with a speculum the two other passageways. If one has good eyesight, one is able to see the mouth of the womb with its cleft, and one will decide rather easily whether it has been opened or not. For once it has been dilated it never closes enough so that a trace of its having been opened cannot be noticed. But for surer confirmation, put a candle to it: if it goes in easily, the passageway has been cleared. It is not at all similar to what we were saying about the large canal, fleshy and soft. This neck is hard, and its resistance is somewhere between that of muscle and cartilage. Once it has given way and relaxed, it is afterward always a little bit open, except when the woman is pregnant. For then, since the entire womb bears down upon the child, the neck of the womb is tight and constrained.

The above are visual and manual tests.[44] I now come to the others, more decent and discreet, but not as certain. Put into the aforementioned valves, with the aid of a funnel, some perfume of jet; or put a small amount of the oil of this substance into the maiden's privities. If you smell the odor of it in her mouth or in her nose from the air she exhales, there is a great likelihood and probability that her back cloister is open. Yet she could well have such a thick womb that the odor would not rise upward, even if the mouth of her womb were opened, as is the case with many a woman, according to the test Hippocrates gave in aphorism 59 of the fifth book.[45]

These are what I see as the signs of virginity, and they are rather unreliable for the reasons I have discussed. I would more willingly rely upon the practice of the Negroes, as told by Pietro Bembo in the sixth book of the *History of Venice* telling of Portuguese navigation.[46] Their midwives, after cutting the umbilical cord, then sew up the first valve, gate, or entryway into the large canal. The girl pisses easily through the stitches, through which the blood of her menses can also flow, but she cannot practice lechery with the boys. Then, when she is married off, on the day of her wedding a little knife is solemnly given to the husband so that he can cut the stitches himself and see for certain whether the entryway has up until then remained closed. For it is not believable that maidens would be so shameless and lascivious, just to take a little free pleasure from it, as to want to be unstitched only to suffer being sewn up again afterward when it came time to be wed.

Yet, I must say, there is a remedy for everything: *& fatta le legge, fatto l'inganno*, as the Italians say.[47] One could do as one does with the ear lobes that are pierced for wearing earrings. Once cicatrized throughout, the hole can take again and again whatever one wishes to put in it without any pain. Thus, too, could some foolish maiden do with her privities, the edges of which are made of the same skin as the ear lobes or the man's prepuce. In antiquity they practiced infibulation, or buckling, as Celsus tells of it, so that boys would not abuse women before they were of age: "One draws back the prepuce," he says, "in the end of which one passes a threaded needle. The thread is left in and is moved every day to keep open the holes until a light scar is formed all the way through. Then a loop is inserted, which can be removed and put back in painlessly."[48] In like manner, one buckles up mares with several rings. A maiden who has been sewn since birth could do just the same. It is a matter of keeping open the holes that were made there, so as to sew herself up again and again at will and practice lechery, even have children, while waiting for a husband. For when her wedding day came, she would not fail to sew herself back up again gently, without hurting herself, just as one laces up a girdle. And her husband (if she wishes) will find the same thread with which she was sewn at her birth, or a similar one, well bepissed and dirtied for the purpose.

And so there is a way for those who have a will to do evil to get around everything, and he brags in vain (as the common proverb goes) who brags about the animal with two holes under its tail. There is for certain one of them that is most difficult—if not impossible—to keep watch over if the goodness, modesty, and decency of the maiden or woman do not. From under the one hundred eyes of Argus, ordained to watch over a heifer, a way was found to remove the hindrance.[49] I do not know if for such a condition one could find a surer remedy than the ring of Hans Carvel, about which Pantagruel will instruct you, if you wish.[50]

CHAPTER V
WHENCE COMES THE COMMUNICATION THAT IS SO EVIDENT BETWEEN THE BREASTS AND THE WOMB

Galen, in Book 14 of *On the Utility of the Parts,* teaches us that the womb and the breasts have common veins, not continuous but adjacent, that can both supply and receive, just as the branchings of the vena porta and the vena cava do with respect to the liver. Vesalius seems to be of the same opinion in chapter 18 of Book 5: "That which is superfluous, and was collected in the veins of the womb, spills over elsewhere in search of a fitting place for movement. The most fitting place is the veins rising along the principal muscle of the abdomen and approaching the veins that descend under the breastbone. For the former empty their blood into the latter, thus accounting for milk being the twin brother of the menses, as the divine Hippocrates has said."

This passage is transcribed from Galen almost word for word. It contradicts not only reason but also visual evidence, because the veins that pass under the pectoral bone and go to the upper part of the principal muscles for their supply of nourishment (as we shall show later) are not adjacent enough to those rising along these muscles to be in contact, as are those of the vena cava and the vena porta in the liver. There are sometimes two large fingers of distance between the ends and orifices of these veins. It is thus apparent that the so-called communication of blood cannot be completed through these particular vessels, which would at least have to touch each other.

In fact, their only purpose is to supply nourishment to the principal muscles, the upper part of which is supplied by the branches of the vein descending under the pectoral bone. Otherwise, why would animals, which do not have breasts on their chests but on the lower belly, have similar veins? Why does a man, who does not have a womb, have them the same as a woman? This proves that they have another purpose than that claimed by common anatomists, since one cannot allege communication between womb and breasts in males.

What, then, is this communication between breasts and the womb, felt in a thousand ways? For if one puts a cupping glass[1] under the breasts, the blood that flows through the womb is retained; and when we wish to reduce the abundance of a woman's milk, we draw blood toward the womb. And it has certainly been observed from time immemorial that milk and the flowers cannot be abundant at the same time, or if so, it is a most rare thing. Because of this it is conjectured that these organs are not only of a common substance but also share some vessels in common. Yet one sees no continuity in the veins coming from these organs unless it be furnished by the vena cava, common to all members, and through which, not without long and convoluted meanderings, the blood can flow from the womb to the breasts, and vice versa.[2] And so we must find some explanation showing us more closely the adjoining and necessary cause of such an effect, which I shall furnish as follows.

Nature, in first configuring the organs, had some of them joined together in a close intimacy, beyond the normal communication existing among all of them, just as she has made elsewhere certain accords and discords, called in Greek sympathies and antipathies. Now, this communication, or mutual accord, is not due to any reason or decision, but only to an inclination and necessity ordered by nature and lying in the very form of things, just as heavy objects fall to the ground and always seek low places because they are of such a constitution and manner that they cannot without violence come to rest elsewhere. In such a way (in my opinion) nature has made the breasts and the womb communicate with a certain intimacy, just as the orifice of the stomach and the diaphragm do with the brain, yet in a more peculiar way, which we are now in search of.

Of the sympathy between the breasts and the womb we have plenty of evidence and many solid arguments. First, when the nipples are tickled, the womb takes delight in it and feels a pleasant titillation. Also, this small button on the breast is highly sensitive because of an abundance of nerve endings; and this is so that, even here, the nipples will have an affinity with the reproductive organs. For just as in these latter organs nature has set some lasciviousness so that animals, driven by voluptuousness, might be inclined to copulate in order to continue their species, so, too, has she done with the breasts, especially in their little buttons, so the female will offer and give suck to the child, who tickles them and pulls on them gently with its tongue and tender mouth. In this the woman can only feel great pleasure, especially when they are abundantly filled with milk.

But what clearer evidence of their alliance can be found than to see them, both together, swell and subside? The ends begin to stand up, and (following the Greek word) itch, which in Languedoc is called *vertilhar*,[3] when the menstrual blood begins to dilate the veins of the womb (which also at that time gets bigger and becomes capable of conceiving). Thus do these organs communicate, so that when a woman is ready to be fecundated, the breasts, being irrigated with the menses and made able to produce lots of milk, are likewise ready to nourish the little infant.

After the woman has conceived, as the child is growing and the womb dilating, the breasts do the same. When the child is brought forth they suddenly receive what was set aside for its nourishment. Later in life, when women lose their flowers (and thus are no longer able to conceive), the womb and breasts shrink little by little and become as small as they were before puberty.

These clear and obvious communications cannot be doubted in any way. Nature has coordinated the breasts and womb to supply the child with nourishment, both upon conception and upon birth, through this same blood, which is more copious than is needed by the mother alone. Sometimes it is the womb that enjoys this blood, and sometimes it is the breasts, according to the needs of the child.

As for the distance between these organs, which seems to work against this communication, it is not as great as one might think. The blood supplying the breasts has not been all the way to the womb, nor has the blood that turns toward the womb been in the breasts. It is, rather, blood that has been in the great vein (which is between the two) and that indifferently flows one way or the other according to how powerfully it is drawn or repelled.

Now, especially influential in this is the emptiness and sponginess of the breasts and the easy dilation of the veins of the womb. For when blood in the trunk of the great vein is too copious, it is shunted into places quick to receive it. It is easily received by the vessels in the womb, which dilate readily. In addition, the womb is located down below, where humors are inclined to settle because of their heaviness. The womb is feeble by nature, in that it was the last to be formed, as its Greek name indicates: *hystere*.[4] If the blood is not driven toward the womb, it is attracted by the breasts, which, nourishing themselves on it first, then make milk from what is left. And they do not cease drawing blood to themselves as long as allowed to do so, for, being spongy, they are able to absorb much more than they would ordinarily. And because either the breasts or the womb harbors all of the superfluous blood, nature forgets one and maintains the other. Thus it happens that blood will be carried and given to the breasts for a long time without going toward the womb, and vice versa, unless there is a huge abundance that can supply both places.

From these arguments one can conclude that the blood abounding in the great vein is sent sometimes to the breasts and sometimes to the womb, according to nature's requirements. She has ordered such communication between these organs so that, in their function of providing sustenance to the child, either the breasts or the womb will make use of the copious blood supply.

There remains one thing to add to the remarks made above. How can a cupping glass placed under the breasts stop the menstrual blood from flowing if there is no communication either through the external veins in the principal muscle or through a natural connection between the breasts and the womb? I reply by saying that the vein rising along the principal muscle branches off from the large vessel going into the womb. Thus, it can easily happen that this vein, exhausted by the cupping glass, will draw upon blood in the veins of the womb and consequently interrupt and suspend the immoderate flow.

CHAPTER VI
WHY IT IS THAT THE MILK FROM A WOMAN WHO HAD A SON IS BETTER FOR NURSING A DAUGHTER, AND VICE VERSA

A lot of people maintain that the milk of a woman who had a son is always better for nursing children of either sex. Our women in Montpellier, held to be more learned than the average (inasmuch as they are in the very fountain of medicine),[1] have a saying, passed down from generation to generation: the milk of a woman who had a daughter is better for a son because (so they say) it cools him; and vice versa, the milk of a woman who had a son is better for a daughter in that it cools her also.

Their statement is on the whole supportable, as we shall demonstrate, but it fails in the arguing. For to allege cooling to both sexes, and to both kinds of milk, is not reasonable. They seek to establish a difference between them but do not set one up, since, according to their argument, all milk cools and both the daughter and the son need to be cooled. This is obviously false, for the male is hotter and the female colder. Thus, if the son must be made colder or cooler so as to temper his constitution, the daughter, on the contrary, must be made hotter rather than cooler to correct her distemperature.

This is why it would be necessary to argue this point differently, and to say that the milk of the woman who had a daughter is better for a son because it cools; and that of a son, for a daughter, so as to make her hotter.[2] But I take it in just the opposite way, claiming that the milk of the woman who had a son is less hot than the the milk of one who had a daughter, and that the daughter needs a milk that is less hot, as I shall easily demonstrate while confirming what our women have learned very well, and which fails to convince only because they have argued it so poorly.

It is first of all necessary to know that every well-complexioned body must be maintained in that particular complexion, and that such maintenance be effected with things of similar quality. This is why nature has instilled an instinct in every body and in each of its parts, down to the very smallest, to draw unto itself the nourishment that is most fitting and that corresponds to its temperature. From several different plants that are in the same region, one will draw from the soil a different sap than will another, and in the same tree different parts draw to themselves different portions of the sap than others, and different parts draw to themselves different portions of the sap in the roots, since the wood is nourished by a different substance than that nourishing the leaves, and the fruit by a different one than that nourishing the bark. It is the same with animals. And in the species of man, more differences are to be found than in all the other species because of the infinitely diverse complexions, as I have already pointed out in the second chapter of the third book.

Among the organs of our bodies or of the other animals, the hottest ones attract and draw for their nourishment and proper upkeep that portion of ordinary blood that is more bilious. The least hot and moist organs attract the pituitous, the driest attract the melancholic. Something similar is done, we must believe, by the child in its mother's womb. If it is a male, since its natural complexion is hotter, it will seek and draw to itself, from the blood it is allotted, that portion that is the most like its complexion. Similarly, the daughter, who is naturally colder, delights in and consequently feeds upon that part of the blood that is less hot than that taken by the son. It follows, therefore, that after birth, in the mother's blood that remains and goes to the breasts to be converted into milk, there will be cooler portions when the child is a boy and hotter portions when it is a girl. For such portions, because they correspond less to the nature of the child, were left behind and spurned as long as the child was able to find elements that it liked better.

Thus, it follows that the milk made from what was left behind by a boy is going to be less hot than that made from what is left behind by a girl. As proof of this one has only to consider the color and consistency of the milk. That which is from a girl[3] is reddish yellow, clear and watery, or serous, poisonlike, a bilious and hot excrement. From a boy, the milk is whiter and thicker, a sign that the heat is far less present in it. And so, the milk of a woman who had a son will be much more fitting for a girl because it is less hot, and because the natural complexion of a girl requires (so as to be maintained in it, according to the nature of her sex) similar nourishment. The son, on the other hand, will be better nourished by the milk of a woman who had a daughter.

There, then, is the justification of what these women say, although for reasons they do not understand. For it is not necessary, properly speaking, to cool either males or females if they are in good health and are born with the temperature that is required by their sex, as we suppose. But the heat of the son must be maintained, as must the lukewarmness of the daughter; otherwise, their natures will be corrupted most inappropriately, making the girl mannish and the boy effeminate.

I already hear an objection being murmured here.[4] "Master, you cried out so much in the first chapter of this book against women who did not nurse their children, and now you prove that the milk of another woman is better for the child than that of its own mother. This is a necessary conclusion based on what you say: the best milk for a son is from a woman who had a daughter, and vice versa. It of course follows that no mother should nurse her child, but that one woman will nurse the son of another, who will nurse the daughter of the first one."

I respond by saying that there is no contradiction in my argument. For I am supposing in this chapter that the mother is unable to nurse, or is legitimately excused from doing so, and thus forced to have recourse to a stranger. In which case I say, and grant, that if one has the choice among wet nurses, the observation of our women is sound: that to a son is given a woman who had a daughter, and vice versa. One might reply[5] by asking why it is not better, since the milk from another is superior, for the mother to have her child nursed by another woman, with the understanding (if you will) that she take someone else's child to nurse so she will not be accused of wanting to preserve her breasts and play the delicate one. But that is not a good argument, since the mother is not expected to return the favor to the one nursing her child as long as she has the means to pay for this service with other compensation.

The main point revolves around whether or not the child would be better nourished by somebody other than its own mother. I say no, and yet I am not contradicting in any way the present argument. For the difference in milk, which we have discussed, is not so great as to make it necessary to prefer this good quality to the quality of maternal milk, which is much better for the child than what was only somewhat better, because it is more familiar to the child and (as Hippocrates says) brother of the menstrual blood that nourished the child in its mother's womb. And, as the same author says of all aliments in general: "Foods and drinks of lesser value but more tasty are to be preferred to better ones that are less enjoyable."[6] Now, one of the things that makes food more tasty is one's being accustomed to it. This is why a mother's milk will always be more proper for the child than will that of another woman, provided the mother is otherwise healthy and not, in fact, ill and considerably out of sorts. For one sees a number of women who are merely sickly and who nurse beautiful children, in spite of their infirmity and weakness. I know there are several mothers who excuse themselves on account of some slight indisposition and give us to believe that their children will not have long to live if they nurse them. It is quite true that good milk is necessary for the nourishment of children, but I am simply saying that if it is only a little bad, it is better coming from the mother than other milk that is only a little better. Thus, one can see how slight is the importance of choosing a nurse who had a son for the nursing of one's daughter (or vice versa), and how great is the importance of preferring the mother's milk for her own fruit, male or female.

To finish this argument I would like to add a small observation made by our physicians, which is almost identical to the one mentioned above. They maintain that if a more cooling or less hot milk is sought, then that of a woman who is nursing a girl is to be prescribed. But in this they are mistaken, in my opinion. First of all, because the boy or girl nursing does not change the kind of milk. For as it is, that is how the milk will stay, whether it is a boy or a girl using it. And so it would be better to ask to have milk from a woman who had a daughter. For (following what I have shown) milk is completely different according to the sex of the child that the woman has carried, not that of the child who suckles. And one could thus refute the argument that when one asks for milk from a woman nursing a daughter, one claims and supposes that such milk comes from the mother herself, following the laws of nature. There would be yet more to say if our first claim is true; for the milk of a woman who carried a daughter is hotter than from a male.[7] And so one is still in error if one asks for this milk, thinking it is more cooling.

CHAPTER VII
THE SUPERSTITIOUS AND FALSE IDEA OF WOMEN WHO BELIEVE THAT THE MILK GLANDS DRY UP ON WOMEN WHOSE MILK IS HEATED

It will not take a long time to refute this notion, among the most absurd and inept of errors, as I shall now demonstrate through true examples and certified experience. I wish to concentrate more on an explanation of what gave rise to such a popular adage. As for the falseness of the claim, it is too obvious, for the same thing is said of goats, sheep, and cows, and yet one sees every day that the milk glands of animals do not dry up when their milk is taken to make porridge. People worthy of trust assured me that, when they were in Nîmes, a woman of the region had such a copious supply of milk that she made porridge out of it for her child to nourish it better, and the more she took from her breasts the more there was to take.

This is a long way from losing it, and boiling it is a lot more than simply warming it. Besides, we often see wet nurses who furnish apothecaries and barbers with some of their milk for remedies that are warmed, and the milk does not dry up in the milk glands of these nurses. Of course, when milk is borrowed from them, they will always say: "Don't you dare heat it." Our people are quick to promise that they will not; yet, knowing that doing so will bring no harm to the nurses, they heat it anyway if there is need to do so, and the nurses lose nothing because of it, thank God.

But where does this idea and the popular adage come from (for few are such sayings that do not have some true sense hidden in them). It is to the wet nurses themselves, and not to their milk, that these words apply, meaning they must keep from getting aroused[1] because that is what makes their milk glands dry up. And "aroused" is to be understood in two ways, mainly. One is the fits of anger and spite to which wet nurses are often subject. For they become proud and haughty because of the great need one has for them, and this forces one's indulgence more than in the case of other servants, out of love for the child. Thus, if one causes them the slightest displeasure, they become crazed and enraged. I mean for the most part, for there are some who are fairly calm and modest. Now, anger or any other violent emotion will heat the humors and will often make the menses flow outside their normal time, causing the material for milk to withdraw. Other times, even if the menses are not provoked, the milk can stop merely through the seething caused by anger, drying up altogether. For the blood that used to be attracted by the breasts goes elsewhere, and once turned away from them, it will not easily return. And so, milk heated because of anger dries up.

The other manner of heating is through love. Here, mothers who give their children up to be nursed have very mistaken ideas, which I shall reveal. Namely, if the nurse is married, these mothers do not want the nurse's husband to know her carnally at all, for fear that he might trouble her milk. They are not entirely wrong in this, but they are also far from being entirely right, for it is much better for the nurse to have the company of her husband, properly and modestly, than for her to burn with love.[2] Violent desire that goes unsatisfied is the main cause of troubled milk, as is seen in excessively amorous nurses who run after men like bitches in heat. Would it not be better to let them have some relief from this great hunger than to make them simmer on a low fire? Sometimes you will see them so troubled with amorous passion that they lose their color completely and are unable to eat or sleep. Who would doubt that the milk would then not be troubled also, and that the milk glands would not dry up? It is necessary that a wet nurse be well fed, that she get plenty of sleep, and do only light work. This regimen incites a person to lust for the deeds of the flesh, pricking, exciting, and driving one to lasciviousness. If an idle woman, well fed and in good health, is tempted by such an impulse and then forced to abstain totally, I think her milk will not be the better for it. Rather, it will be troubled, heated, and will smell like a goat, just like her own body. So it would be preferable for her to enjoy her husband moderately, as we said, than for her to be deprived of him and to be completely sequestered.

Stop to think! Are the wives of plowmen, artisans, merchants, and others who commonly nurse their children excluded from their husbands' beds? Or do their husbands keep from making love to them as long as they are nursing? One can see very well that such is not the case. And are their children any less well nourished? Are they more weak or sickly than children of delicate city women, affected gentlewomen, or munificent great ladies, who do not wish to lower themselves to the point of assuming this duty of nature by nursing their children with the milk God has given them so that they can be mothers in a complete sense?[3] Far from it! The children of poor women, nursed by their mothers, are usually stronger and healthier.

But it is feared (here is the biggest reason) that the nurse might become pregnant by sleeping with her husband, and that the child will be sucking bad milk, rendered thus without any doubt because of the pregnancy.[4] And it is to be feared that the nurse will not be aware she is pregnant until the milk has been found to be quite bad. For most women do not have their menses while they are nursing and thus do not realize they are pregnant until their milk starts to fail. Others who go on having their flowers are often one month pregnant before they learn of it. What is worse, there are wet nurses who, knowing full well they are pregnant, say nothing about it as long as they have a drop of milk left, afraid they will be dismissed. And thus, they abuse the child, called in Languedoc *enganar*, from an Italian word for it: *ingannare*.[5]

These are the main reasons that proper women give for not wanting the nurses of their children to sleep with men. But the disadvantages I presented above counterbalance them considerably and (to my mind) tip the scales, in that they are more weighty. For the overheated milk of a woman passionately in love is by far worse and more harmful than that of a pregnant woman.

And after all, is it not true (as we said in the second chapter of this book) that village women have no trouble nursing their children, even if they feel they are pregnant, as long as there is a drop of milk in their breasts and the child is able to suck it out? If they could nurse until the ninth month, they would continue with no difficulty and then wean it after it is a year old. Are they more wretched and inept at their work? It is clear that they are more robust and patient in their toil than city dwellers. Poor people say that if the child has drunk the best of this liquid, it must also in the end drink the dregs, just as they themselves do with their wine. For they drink both the bottom and the top, as long as the spigot flows, right down to the last drop. But people who are soft and sickly, well-off and dainty, stop drinking the wine as soon as it gets down to half a barrel, saying it smells of the bottom. Servants and chambermaids drink the rest, right down to the dregs. So, too, will it be with children who are nursed, whose wine is milk, just as we say, conversely, that the milk of the elderly is wine. Thus, the above comparison is most fitting.

The ladies who will understand these words incorrectly will say that I advise the nursing of children with milk from a pregnant woman. But (with all due respect) that is not what I am giving as advice, but what I am giving is an example to show how, for children of poor village folks, coarsely nourished, the milk of their pregnant mothers is not harmful to them.[6] I am not saying it will do no harm to delicate children from good houses, because they are from parents who were daintily nourished and because it was not with milk from their mothers. For it must be understood that there is such an affinity between the child and the blood of its mother that it would be better off fed with the worst milk its mother could have than the best of some other woman. I know well that this statement will seem strange, but it is true, and I shall prove it well enough in the sixth book, which will be on custom.[7]

And if I have accomplished nothing other than convincing people that milk from a pregnant woman is not as bad for the child as that of a woman who is as hot as a bitch, excessively desirous of the company of her husband or friend, I have sufficiently saved from error those who find it so unusual that a wet nurse enjoys her love-making. It is always to be understood that it be in moderation and soberly, as one always does when enjoying full liberty. For if it has to be done secretly and stealthily, one goes about it like an unsaddled donkey, getting so heated up that double evils result: one is that the milk is greatly troubled by it; the other is that the wet nurses get pregnant more quickly in this manner. For it is as if one were to keep a drunkard's wine locked up; if he finds the key to the cellar he will drink as much as he can. Let him have as much wine as he wants, when he wants it, and he will drink far less by a considerable amount and, as a result, will be sober more often.

"Many thanks," the wet nurses will say when they hear this. "You have spoken very well on our behalf. This is a good recipe. We will follow it willingly. You are a good doctor. May God keep you from harm." And the mistresses, on the contrary, will think I am in love with the wet nurses, and that I enjoy caressing them. It is certainly true that I love wet nurses, and that the woman in this world that I cherish the most has nursed all my children for as long as she had milk. I never stopped sleeping with her because of it nor stopped making love to her, as a good other half to hers, performing the conjunction of marriage. Our children (thank God) were well nourished and turned out well. I give no advice to others that I do not follow myself.

There, then, is what is to be understood when ignorant people claim that the heating of the milk causes wet nurses' milk glands to dry up. There is also another way of understanding what is said about it in animals: that they dry up not if one simply boils the milk (such as when one makes porridge out of it), but if it spills over into the fire, as can happen when it boils vigorously. Likewise, if some water is not added to it, common people say (at least in Gascony, where I picked it up) that the milk glands of the animal will dry up.

There are mysteries or secrets in these words: one is an encouragement to be parsimonious, or frugal; the other is an admonition to heat milk properly. As for the first, it is good advice indeed not to allow milk to spill into the fire or anywhere else. For if one clumsily spills some, one could run out of it, and the breasts providing it will go dry, that is, will not be able to produce enough. For that very reason it is good to supplement it with a little water, so that less milk will suffice; otherwise, it will come to be lacking, or more animals will be needed to provide it. Thus, it seems that the animal is going dry when it is unable to provide all that is needed from it.

As for the second, it is a good precept for secretly telling how milk ought to be heated. It must be at low heat because the substance in it, being very unstable, will not stand such vigorous heat as would cause it to boil up and spill over. For that very reason it is very good to add to it a little water, which resists, and causes the milk to resist, the adustion of the fire.[8] Thus, at the same time it will be heated more gently, and less of it will be needed. These are the two hidden reasons behind the belief instilled in people of humble means so they will be better able to make their milk last and to heat it better. For one could find no better way of gently getting them to do something than by threatening them with considerable harm or, conversely, enticing them with great gain.

CHAPTER VIII
THAT NIPPLES MUST NOT BE HARDENED IN ORDER TO AVOID *TANDRIERES*[1]

Tandrieres are little fissures in the ends or nipples of the breasts, when they crack and split because of the first milk, especially in those who nurse, because as the child sucks and pulls on them, they split all the more. This happens mainly to more delicate, soft, and dainty women, in whom the ailment is called *tandrieres*. Once the teat has been split and hardens, a woman does not have much pain (or very little) in future pregnancies.

Now, in order to avoid them, especially right from the first child, women apply diverse remedies, all of which tend toward exsiccation, thinking that by correcting the softness one will prevent such little splits, inasmuch as the nipple that is already hardened is not going to be so subject to them. For this reason, some bathe their nipples with water and alum; others, with rose water and plantain, or with myrtle or some other astringent. And all these do nothing but make the nipples much worse, for the harder and stiffer the nipples are, the more they will split.

The very opposite is what must be done: soften them and make them supple before the milk arrives. For if they are soft, they will definitely give enough and will not split. Just as our lips, which chap and crack in the winter because of the drying and stiffening cold, are preserved from this condition if one wets and rewets them with saliva, or if one puts pomatum[2] on them. This is why those who are better informed will apply to their nipples some new wax thinned with some soft oil a few months before delivering in order to avoid *tandrieres*. But it is better yet, as I prescribe, to grease them often with cool bacon fat, which softens them nicely and gently. The reason for it is simple, and the experience of several women confirms it. I base my claim on the word of those I taught to use this remedy; they found that it worked very well.[3] I thought I would mention it here for those who, concerned about nursing their children, might excuse themselves from doing so only on account of this particular ailment. I am not at all concerned about the others, who care little for their children and are disdainful of nursing them.

CHAPTER IX
ON CHANGING THE CHILD WHEN IT IS DIRTY; AND WHETHER IT SHOULD HAVE FIXED HOURS FOR FEEDINGS

Women are of the opinion that in order to give a child proper care it must be regulated according to fixed hours, both for nursing as well as for changing its breechcloth to keep it clean. And this "proper care," as they call it, is usually understood to mean a convenient caring that will not cause much trouble for its mother or nurse once the child has been regimented and accustomed to a certain regimen and hourly routine corresponding to the convenience of the woman who nurses it. Thus, such a regimen conforms more to the nurse than to the child. If one manages to force the child into this mold it is said to be good-natured. That is, it does not demand anything at an inopportune time, but only during one of the moments set aside for it.

But let us see whether this regimen and strict regulation are beneficial for the child, first in the area of nursing, and second in that of its breechcloth. For if the child normally suckles at fixed times, it will also void regularly, as long as there is no upset in the stomach and the child is in good health, and as long as the milk continues to be the same, not becoming watery, thick, acrid, or sharp.[1] For these diverse qualities will quickly change the child's stomach.

Let us consider first whether it is good and beneficial for the child to nurse only at fixed times. We pointed out in the second chapter of this book that the child in the womb draws its nourishment continuously from its mother through the umbilical cord, as a plant ceaselessly draws food from the earth through its roots. Once born, breathing the air and taking its sustenance by mouth, it needs to be fed frequently, especially since its body, soft and tender as cheese (to which Galen has compared it), changes and evolves constantly. Thus, if what is dissipated at every moment is not replenished through frequent feedings, the child will remain small, weakened, and withered. And the high frequency[2] of the feedings is all the more necessary in the first few days because the child is closer to the time when it was continuously drawing in its food. This is why, so as not to have a sudden change from one extreme to another (a thing unbearable for nature), frequency of the feedings must correspond to the continuous ingestion that the child used to enjoy.

Also, its stomach is so small that it cannot hold much at one time until it is stretched, which comes about little by little. Thus, it is necessary that, in the meantime, frequent repetition make up for the small quantity of food. Later, when its stomach is larger, the child has less need of suckling as often as before, because its body is proportionately larger, and it needs more food than during the first few days. And so, it is always necessary for the child to keep suckling often until it begins to eat something. For then, being nourished with a more substantial food than milk, its stomach is slower at digesting and does not require feedings as frequently as before.

All of this will be granted me, but the main point is yet to be seen: whether or not one ought, or is able without harm or risk to the child, to limit and restrict its desire to nurse to certain hours so as to have fixed times, such as every two, three, or four hours, and so on, at whatever intervals one might advise.

The women of Montpellier usually set their times at every four hours, which allows for a feeding six times in a complete day, counting day and night. This seems reasonable enough, but it is impossible to put all children in the same category, since all are not of the same complexion and nature. It is well known that with large as well as small children some are very hungry and others are not. The latter will go for a long time without nursing, the former want to have the nipple in their mouths nearly every minute, and if they are refused, or if they are not often offered the breast, they become undernourished. The size and capacity of the stomach are different in different bodies right from the first development. There are small and large livers, small and large heads, little hands and long fingers, and so on and so forth with the other parts, which do not always match the rest of the body. A large body will sometimes have a very small stomach, and a small body a large one. It thus often happens that a child of considerable corpulence will need to suckle every hour, because the stomach is small and the body needs a lot of nourishment. Its small stomach can scarcely hold it all at once, and if it calls for a lot of food, stimulated by the needs of the other organs, it is forced to reject and vomit this milk, more copious than the stomach can properly absorb. There are, on the other hand, small and wretched children who absorb milk like a sponge and swallow it like an abyss because they have huge and capacious stomachs. They will have enough for several hours with just one feeding.

So, whoever would seek to limit the feedings of all children to the same times cannot fail upsetting nearly all of them. People are willing to grant me this also, but it remains doubtful whether or not children's feedings can be strictly regulated by setting up various limits according to each one's diverse complexion and nature, which can be discovered in a few days. I will tell you: if the nurse is so watchful, concerned, and intelligent as to understand clearly the condition of the child, and so kind as to want to conform to it completely, complying in every way to the times required by the child's nature, no harm is done in allowing her to begin and to stop as she sees fit, and to continue giving it suck at precisely such times. For the child that is fed on an individual basis will be far healthier.

But how many wet nurses will you find, whether mothers or hired women, who will be so concerned and prudently observant as to be able to tell, or who, having once understood it, will not prefer imposing upon the child a routine more convenient for herself than accommodating herself to the child? Who wants to deprive oneself of one's pleasures, recreation, meals, and sleep in order to give oneself totally to the timetable the child requires based upon its own complexion? You will scarcely find ten in a thousand who will be thus disposed.

And so it seems it would be better to make another rule: namely, that the child not have any fixed and limited timetable, but that the nurse give it her breast constantly. For if it is hungry, it will suck; if not, it will abstain. How can one expect to regulate a child, since every time it complains or cries about something (such as a pin that is pricking it or a flea that is biting it) one has to have quick recourse to the breast to appease it. It is therefore often necessary to forget going by fixed and limited times, however upset one might be about it. And if one makes an exception on such occasions without harming the child, neither will it be harmed when it is offered the breast at other occasions or at unfixed times.

But our women fear such subjection and admit it openly. Some are so attached to their own pleasures that they do not want the servant girl to bring them the child when it is crying about something so it can be appeased with the breast, if it is not at the set time. They expect her to carry it about and sing it songs or to rock it back and forth and put it to sleep. But perhaps the child is crying because of hunger. How, then, do you expect to put it to sleep? They know very well how to recite the common proverb: *Qui non ha lou ventre dur, non pot pas dormir segur.*[3] So, the child that has a flat and soft stomach, troubled with hunger before its usual time, will not be able to sleep. To appease it or satisfy it with a song is a pure mockery. I would very much like to know if its nurse, finding herself in good appetite, would be satisfied and content with a song instead of some soup (even if she were Orlando de Lassus),[4] or with dancing the Champagne shake.[5] What foolishness! We say in a Latin proverb: "A hungry stomach has no ears,"[6] and in a verse from days gone by, "An empty stomach will hear no arguments."[7]

"But I have company," the gentlewoman will say. "Do you expect the child to be brought to me here and for me to show my breast?" There we have a great risk indeed, and a most pertinent excuse! I am ashamed of such replies, which I find more repugnant than the matter we are about to treat, for it is time to come to the changing of the child.

On this subject I have already said that if the child were always able to nurse at the same times and if the milk did not change in quality, the child, too, could void at fixed times. But since the first is lacking, so, too, is the second. This is why one cannot have fixed, limited, and determined times for changing a child, which cannot and must not be disregarded in case of a necessity. This would be (to my mind) each and every time one realizes the child has bepissed and beshat itself, even if it has not been one hour ago that it was last completely changed.

And what point is there to making it endure this stench and filth for four or five hours, until its time comes? If a man sweats from hard work, people find it good for him to change shirts immediately and not to soak in this sweat, and even less to let it chill on his body. And how will it be good for the child to have it soak in its urine for four or five hours? What good could that possibly do it? And the same for its shit? The women reply that *"entre la merde & lou pis se nourris lou bel fis."*[8] But I explained this saying better, to tell the truth, in the sixth chapter of the fourth book, and how it is to be understood: that every child is nourished amidst piss and shit, whether beautiful or ugly. And it has no ill effect on beauty. For if they wish to say[9] that these materials are cleansing agents, cleaning the skin and making for a beautiful complexion, and that this is why one wipes the faces of children with the bepissed breechcloths of smaller ones in order to cleanse and beautify them, I answer by saying that children have no need of these cosmetics or beautification on their thighs, legs, belly, loins, and arms, and that there is a huge difference between wiping them with these things and letting them soak in them for four or five hours.

From such soaking there often come about many ills, both of mind and of body, which I want midwives to take careful note of. First, as for the body, they know very well that this filth often causes rashes on the thighs and buttocks of children, causing them to become angry and to cry, not without reason. This is due to the sharpness or burning of these excrements, which often become this way because of the prolonged contact with the child's body, made to endure this Gehenna unjustly. As for the mind, it is doubly offended by the excrement and receives bad sensations from it. One of them has already been mentioned: that the children become angry and cry, which is a bad habit formed by often repeated acts and certain dispositions. For after having been accustomed over a long period of time to crying and screaming over the discomfort this filth causes them, they are so upset afterward that the smallest thing in the world makes them angry. Thus, the mothers and nurses are properly punished for failing to keep the child clean. For afterward they have a much worse time of it when the child becomes ill-tempered for having suffered too much. But I do not pity them as much as I do the poor innocent little one, whose mind has been altered and who will feel it for the rest of his life.

There is another way in which the child is offended by its excrement when one accustoms the child's body to it. It is that since behavior corresponds to the temperature of the body (as we have said often), it naturally follows that in a body nourished in filth and excrement there will be a soul taking delight in every sort of vileness, more than if its body had been gently and neatly cared for. I ask you, see if the cowherds, hogherds, stablemen, chimney sweeps, horsekeepers, privy cleaners, and road workers do not have worse manners and less proper language than other people. One enjoys what one was raised in, for nurture is stronger than nature. Let mothers be warned, then, as well as all nurses in general, not to regret efforts spent in changing children as often as they get dirty, day or night. They will be handsomely rewarded when the children become more manageable, kind, and sweet. On the contrary, for every hour that they spare themselves the trouble, the children will give them over a thousand of the worst kind.

CHAPTER X
AGAINST THOSE WHO THINK IT IS GOOD FOR CHILDREN TO CRY AND SCREAM

From what I have shown in the preceding chapter, it is possible to confound and overturn this error. For even if it were only the mind that became more and more unsound by getting accustomed to screaming and hollering over every little thing, it would be a great evil, since one must always wish (as the ancients used to say) for a sound mind in a sound body.[1] But it is also most harmful to the body of the child for it to be left to cry when it could be appeased. For this could alter little by little its temperature in the direction of hot and dry anger, which will keep the child skinny and slight and even shorten its life span (as we said in the second chapter of the first book). There are some children who become so saddened and morose over the little account taken of their crying that they become completely ill-humored from tormenting themselves. Others become short of breath and almost stifle. There are some who become pale, as if they were dead. Some have fainting spells over it, others burst open[2] and then must be castrated.

These are some of the great misfortunes that happen fairly often because of the little account taken of children's crying. And as for the gains and advantages, I do not know of any, except for the enlargement of the lungs, chest, and heart (as our wives say, *"Lou coret luy crés"*),[3] the increase of natural heat, or the burning of superfluities (just as it is said of weeping that it flushes the brain).

I do not find weeping a bad thing, provided it is a medium crying and not excessive, just as medium cries that are neither deceitful nor extreme do not seem to me to be at all harmful to children's health. It is so much exercise for them, so to speak, and they gain from it what was mentioned just above.

But getting used to it is always bad, for it is easy to go from medium to excessive crying. And what woman is there in the world who would not find it a good thing if a child never cried but was always peaceful, happy, joyous, and jovial? I think there would not be one woman who would want to provoke the child to cry and sob, saying it was better for it. But if the child happens to cry, and the nurse (whether the mother or a hireling) does not have the time or the desire to appease it immediately, she will justify not tending to it by saying that crying and screaming do it a lot of good.

This is how one flatters and spares oneself a great deal of trouble most inappropriately, to the detriment of the child, who in the long run suffers from this rigor, austerity, and cruelty, masked and disguised under a sound reason. I say that it can be recognized as much in the body as in the mind of the child, and I dare say that children thus poorly cared for never will love their mothers and nurses as much as if they had been raised with more tenderness. For it is here that compassion and charitable love must begin, which God later makes reciprocal between children and parents. The baby stork gives such a telling example of this that the Greeks have gone so far as to name this type of gratitude *antipelargie*, after the stork.[4]

I am not advocating in all this an overdone daintiness and excessive indulgence on the part of mothers, especially when children begin to recognize their mothers, for from that point on I raise them under the shadow of the rod and make them fear punishment, even before they are weaned. Otherwise, if one is too afraid of upsetting them, they will not be afraid of reprimands, and one will be forced to be overly subject to them, putting up with all their faults and bad habits. This is also why God allows fathers and mothers to always have their children subject to them. Children should not be provoked to anger and to spitefulness, but neither must one be fearful or subject to their emotions, which ought, on the contrary, to be surely guided little by little through discipline to take firm root and to be kept from going in too many directions.[5] And so, certainly a small amount of crying and weeping can do them no harm, and it is necessary that, whatever might or does happen, they take early on the road to virtue.

CHAPTER XI
WHICH SHOULD BE NURSED LONGER, A SON OR A DAUGHTER; AND HOW LONG FOR EACH

Different countries have different customs, and just as the manner of dress is diverse, so, too, is the manner of living. This is most understandable, for the difference in the air and in the soil causes a different kind of behavior.[1] For example, in the cold northernmost countries, hot rooms, fur coats, stoves, wine, and spices are necessary and normal; and in hot and baked regions, such as that of the Moors, underground places are the best, and going around naked, drinking water, and eating a lot of refreshing fruits. Anybody wanting to live in Africa, Mauretania, or Ethiopia in the same way as one lives in England, Germany, or Poland, or vice versa,[2] could scarcely last if he adopted such behavior.

But there is no need to make comparisons with such farfetched countries. If a Parisian wanted to live as they do in Provence, a person from Lyons as they do in Spain, or a mountain man as do those from the low countries, and vice versa, without ever leaving his native land (that goes without saying), he would not be at ease. The sky or the different air causes different behavior in us, and so does the nature of the people, which we call their way of life.

For if one raised a plowman as one raises a man of letters or some other sedentary person, he would become very delicate and could no longer do his plowing. Conversely, if a sedentary man were raised as a plowman, he would soon fall sick because he would not be able to digest the coarse diet of a plowman—unless he happened to be of a strong and robust constitution, having been born of poor people, plowmen, or artisans and, consequently, coarsely nourished. This is why plowmen are extremely well-conditioned for work and are able to live on almost the same food as did their parents, without suffering from it in the least, as is the case with most men of letters.

Age also causes different behavior, since it is a different complexion. Everywhere in the world it is observed that infants are treated differently from boys, boys differently from grown and adult men, and the elderly differently still, and decrepit people differently yet. Likewise, the two sexes are treated differently, not only in dress but also in food and education. Whence the common saying that the boy ought to be well fed, well beaten, and poorly dressed; and the girl, well dressed, well beaten, and poorly fed.

Now, I shall leave aside the different ways of raising children who are nursed according to the customs of their regions, since it is most necessary that they be nursed differently, just as their regions also differ. I will restrict my words to the climate in Montpellier and in the surrounding regions, corresponding rather closely to the temperature of Tuscany.

Our women believe that girls ought to suckle for a shorter time than boys: eight to ten months is enough for girls; as for the boys, twenty-four months, that is, two whole years. One must always presuppose that the child is in good health and developing normally (according to the level of its age), that it started eating when it should have, that it had enough teeth with which to chew, and that the weaning, so dreaded, turned out well. In short, there must be no other question than that of duration.

The reason that leads women to say that girls should not be nursed as long as boys is (to my mind) because they are more moist. True, but what must be known is whether this moisture is harmful or not. If it is the natural state of the female sex to be more moist, and if nature made her that way on purpose, and colder as well (for the reasons pointed out in the first chapter of the second book), would it not be a bad thing to make girls drier, risking the danger of making them manly and sterile? If it were a superfluous moisture, and one acquired through bad food, in or out of the womb, it would have to be burned off. But this is a natural moisture (assuming, of course, that the girl is born healthy, very strong, and of a sound constitution).

Do you want to make a boy out of her by drying her to the point that she lacks nothing but the virile member? She will even have a beard! It is very poor reasoning to say that the girl ought to be nursed for a shorter time because she is too moist, when in fact she would have to be nursed longer to be maintained in this complexion. For such a complexion is not only natural to her but is necessary in order for her to be fecund and have beautiful children—the perfection of the female sex. She will then have a longer adolescence,[3] one not limited by considerable exsiccation, in which the bones and other solid organs become unable to stretch and lengthen. Another good reason for nursing girls longer is because one is right in nursing sons longer because of their dryness. For if one does not keep at bay as long as possible the great exsiccation (to which boys' natural heat drives them much more strongly than it does girls), it is certain that they will stay short, and with the passage of time, their sons and grandsons will be nothing more than dwarfs.

This is seen quite often in those who were poorly nourished or fed with bad or different milks, or who did not suckle enough. They are considerably smaller than others of the same lineage, house, or condition. Hence, it is not a bad idea to let the son nurse longer in order to have beautiful men who will also live longer, according to the course of nature, and get older later in life. For aging is nothing more than drying out, and natural death is but an extreme exsiccation, which can be delayed if at every age one is careful to save and properly maintain one's natural and radical humor, in which the certain measure and duration of our lives resides, as we have amply shown in the second chapter of the first book.

But why will it not be a good thing for a girl to be nursed for as long a time,[4] given the above reasons, which seem common to both sexes? If the mother of one and the other is very healthy, neither phlegmatic nor catarrhal, and if the children have exactly the complexion required for their sexes, it seems to me that one should not treat them differently in any way, even after what we pointed out in the fifth chapter of this book, where we showed that the complexion of each sex must be maintained naturally by like humors. This is why the milk of a woman who had a son is better for a girl, because it is more cold and moist, against popular opinion.

Why then, do laymen think that a girl has less need of nursing than a boy? I think they picked this notion up from some learned physicians without understanding it fully. Moreover, because these laymen think it out poorly, they postulate a cause that does not exist. Laymen also reason poorly when they say that the milk of a woman who had a son is better for a girl, and vice versa, when they say it will cool them.

What, then, is the real reason? It is this, in my judgment: the ancient physicians, who might have spoken of this matter to ordinary people, always presupposed that each mother would do her duty of nursing her children. Now, the milk of a woman who had a son is cooler and more moist. This is why it reduces the child's natural heat and dryness. This is in no way dangerous for its temperature or complexion; on the contrary, it yields a certain benefit: the child will keep growing for a longer time and become larger. And so, there is no danger if the son is nursed for a considerable time, and he should be nursed even longer if the milk matches his complexion perfectly.

Likewise, the girl who sucks her mother's milk, which is hotter and drier, will be somewhat upset in her constitution and can even be so altered little by little that her body will not keep growing as much as it would by using similar milk. Thus, it is better if she is weaned sooner.

"But wait," someone will say.[5] "Is not the food she is given afterward in place of the milk more desiccative than the milk that was taken away from her?" It is certain that the milk moistens considerably, being a benign food that is easy to digest and very nourishing; but it is hotter than blood, and meat is made of blood. Thus, the meat of animals, which we eat (and even more so the broth made from it), yields less heat than does milk.[6] As proof thereof we recall that if the nurse is the slightest bit angry, or otherwise overheated, her child (if it is sensitive) will soon be all stirred up, red, and serpiginous. This is because the milk has such a temper that only a little more heat will make it as strong as wine, to which it is also similar. For both one and the other are nourishing,[7] hot, and moist, as foods; still, wine is hotter, unless it is diluted with water, and then it corresponds to the temperature of milk.

I know that several people will be upset by my maintaining that milk is hot. For it is commonly said that it is made from blood that has been rendered less refined or coarse again in the breasts. This I pertinently deny. For it is made from blood that is worked and elaborated upon in the milk glands of the breasts, which are hotter than they are cold, as I maintain all the spermatic parts to be. But this difference is matter for our schools of medicine.

There remains the fact that the food given to the child after it is weaned is not as hot as milk, unless one gives it some insufficiently diluted wine. But meat and broth, which are fairly moistening, do not heat up the child at all (except in that they are food) and are more substantial. Thus, they make the child stronger. Also, one sees that those who were nursed the longest are for the most part soft, sensitive, and effeminate. It is unavoidable for the first few days that the child be fed with milk, for three main reasons. One is that every change must be made little by little; and there is not much difference between the blood that nourished it in the womb and the milk that was since made from it. Another is that the child has a natural inclination to suckle and knows how to do it without being taught; so, it latches onto the nipple better than it would be able to swallow broth. But the first reason is more valid. Add to these a third: milk is easier to digest than broth, meat, bread, and other foods; and the child's soft and tender stomach is unable to take in anything but moist and temperately warm milk.

Now, onward! All that is granted. Let us now cease and conclude by saying how long a boy and a girl should be nursed. I said that an equal period is due to both if one can choose the milk; that is, that the milk from a woman who had a son be given to the girl, and vice versa. Otherwise, if the milk one gives a girl is for a male child, it is better to wean her sooner, at a year and a half; and to let the boy be nursed with any milk until he is two years old, provided it is high in nourishment. I see no danger in this whatsoever.

END OF THE FIFTH BOOK AND END OF PART ONE

APPENDIX A
DIVISION OF THE ENTIRE WORK INTO SIX PARTS CONTAINING THIRTY BOOKS

FIRST PART

Concerning Medicine and Physicians.	Book I.
Concerning the Venereal Act, Conception, and Generation.	Book II.
Concerning Pregnancy.	Book III.
Concerning Childbirth and Lying-in.	Book IV.
Concerning Milk and the Child's Nourishment.	Book V.

SECOND PART

Concerning Complexion and Custom.	Book VI.
Concerning Height and Weight.	Book VII.
Concerning Air and Clothes.	Book VIII.

Concerning Appetite and Thirst.	Book IX.
Concerning Meals and Digestion.	Book X.

THIRD PART

Concerning Eating and Food.	Book XI.
Concerning Preparation and Order in Serving Food.	Book XII.
Concerning Fruits, Salads, and Cheese.	Book XIII.
Concerning Drink.	Book XIV.
Treatise on Wine.	Book XV.

FOURTH PART

Concerning Retiring and Sleeping.	Book XVI.
Concerning the Causes of Disease.	Book XVII.
Concerning Diseases.	Book XVIII.
Concerning Prognosis in Illnesses.	Book XIX.
Concerning Food during Illnesses.	Book XX.

FIFTH PART

Concerning the Curing of Illnesses.	Book XXI.
Concerning Abuses and Remedies.	Book XXII.
Concerning Poor Treatment and Extravagant Remedies.	Book XXIII.
Concerning Superstitious and Useless Remedies.	Book XXIV.
Concerning Good and True Remedies.	Book XXV.

SIXTH PART

DIVISION OF THE FIRST PART INTO BOOKS AND CHAPTERS[1]

[The Table of Contents of Part One appears on pp. 11–15 and therefore will not be reproduced here.]

POPULAR SOURCES OF THE FIRST FIVE BOOKS[2]

OF THE FIRST

1. Against those who are of the opinion that surgeons are not qualified to reset dislocations, and who want empirical resetters because they are more clever and successful.

2. Why people say that because of a young physician cemeteries are bloated, and that bad physicians arrive on horseback and leave on foot.

3. Against those who say that we will live until we die in spite of physicians.

4. Against those who say: "Piss clear and make a fig at physicians"; and others who say that he who pisses, sleeps, and wags well has no need of Doctor Bell; item, he who has sanicle and bugle can thumb his nose at physicians.

5. That he who thinks himself in good health carries death in his bosom; and whether what one says is true: "Far from a city, far from health."

6. Whether it is folly, as people commonly say, to make your physician your heir.

7. How one ought to understand the saying that the wealth of physicians is completely ill acquired.

8. Why people say that there are more old drunkards than there are old physicians.

9. Against those who say that to live on medication is to live in misery.

10. How one ought to understand the saying that nothing can hurt in the presence of a physician.

11. What is understood by the saying "a physician in tune with nature"; and why people say that a miserable physician makes the wound fester.

12. Whether it is fitting for the physician to deceive the patient.

13. Why people say that physicians no longer travel about on mules since the discovery of scraping and of having chilblains.[3]

OF THE SECOND

1. Whether it is true that truffles, artichokes, and oysters make a man more lively in love's games.

2. Concerning a woman who fed her husband one of his testicles that had been removed, thinking he would be as lusty as before.

3. Concerning charming a man's codpiece, what it means, and how it can be done.[4]

4. Whether it is true that a pale woman requires a male.

5. Whether it is true that women are in danger of becoming dizzy if they are not married off before twenty-eight years of age.

6. Why people say that marriages during the month of May are for the most part unhappy ones.

7. Whether it is true that making love while standing up causes gout, as they say.

8. How it can be that a boy is able to beget at ten years of age, as has been duly witnessed.

9. Whether it is true that a man gets older faster by sleeping with an old woman, and that an old woman gets younger by sleeping with a young man.

10. Whether it is true that nocturnal pollutions would be so many children.

11. Whether it is true that a woman conceives, or retains the semen, if she pisses soon after copulation.

12. Why people say that a drunk woman is as good as pregnant.

13. Whether it is true that men with hernias, or who are ruptured, usually have more children than others.

14. Whence it comes that public whores conceive very rarely.

15. Why it is said that the stupidest people have more children.

16. Whether it is true that women become stronger and men weaker from the venereal act.

17. Whence it comes that several strong, healthy, and vigorous

women cannot conceive, and on the contrary, that several unhealthy ones who are almost always sick have many children.

OF THE THIRD

1. Whether a child of less than seven months can be viable.
2. Concerning pregnant women who drink aquavit as soon as they enter their ninth month so that their child will not be scurfy.
3. Whether a woman who miscarries is in greater danger than one who carries her child to term.

OF THE FOURTH

1. Whether it has any influence on the delivery if a woman in labor says three times while moving her thumb quickly: "I'm hot, I'm cold."
2. Why people say of a man who is lively and lusty that he was born hairy.

OF THE FIFTH

1. Against those who do not want the nipples to be touched either with salves or with iron.
2. Whether it is good to stop children from using their left hands.
3. Whether, in order to strengthen a very weak child, the change to an older milk is necessary.
4. Concerning those who do not want people to get fire in the house of a woman who has just delivered for fear that her child will drool or will be blear-eyed.
5. Why people speak of a child beautiful right down to the teeth.
6. How it can be that a wet nurse is able to know by her nipples that her absent child is crying.
7. Whether it is true that by kissing small children often one drinks up their blood.
8. Whether it is true that a child's buttocks will be excoriated if red coals or hot ashes are thrown on its excrement.
9. What the symbolism is of a nurse's gift of eggs and salt to a child when first brought into the house of a friend.
10. Against those who think it well and good that wet nurses chew food for children.
11. Whether it is true that children who drink goat's milk in summer are in danger of becoming very ill the following autumn.

DIVISION OF THE SECOND PART INTO BOOKS AND CHAPTERS

BOOK VI
CONCERNING COMPLEXION AND CUSTOM

How it should be understood that every seven years one changes one's nature or complexion. Chap. I.

That each should know his complexion and strength, so as to describe it quickly to a physician. Chap. II.

That the physician who knew the patient when he was healthy is more able to cure him. Chap. III.

Whether it is possible for a physician to grasp in a short time a person's complexion; and whether it is necessary to trust completely those who say they have known him for a long time. Chap. IV.

Against those who invoke in every instance their custom, especially when they move on in age. Chap. V.

Whether it is true when one says: "Bad custom and good cake do not mix."[5] Chap. VI.

BOOK VII[6]
CONCERNING HEIGHT AND WEIGHT

Why it is said of those who grow fast that weeds grow quickly. Chap. I.

Whether it is true that stretching the arms and legs a lot each morning before rising will cause one to grow more. Chap. II.

Against those who say that passing one's leg over a child's head stops it from growing. Chap. III.

Whether it is true that garters stop growth, cause wrinkles on the face, and ruin eyesight. Chap. IV.

Why people say that accustoming one to binding from youth hinders or retards growth. Chap. V.

Whether it is true that a child has half of the height it will ever have at age three. Chap. VI.

Whether it is true that one grows while sleeping, and that the work during the day diminishes as much from the growth as one can acquire during sleep. Chap. VII.

Whether the end of the fingers being fleshy means that the person is, or will become, fat; and whether the ends of the fingers being bony is a sign of skinniness. Chap. VIII.

Whether one should eat often and a lot in order to gain weight. Chap. IX.

Which is more fattening and better nourishing, a broth or a roast; and whether it is true that salt and vinegar cause skinniness. Chap. X.

Concerning those who remain standing a long time just after a meal so as not to get fat. Chap. XI.

Guaranteed ways of losing weight, and others for putting it on. Chap. XII.

What is the most advantageous state of a person said to be of a sound disposition. Chap. XIII.

BOOK VIII
CONCERNING AIR AND CLOTHES

Against those who say it is a bad habit to stay inside in winter. Chap. I.

Whether it is true that warming the bed causes scurf. Chap. II.

Whether it is good to feel the cold; and what it is to be well prepared for winter. Chap. III.

Whether it is right to say: "The top, the bottom, and the middle hot, about the rest you should worry not." Chap. IV.

Why people say that the first cold weather is the most dangerous, and the March sun likewise. Chap. V.

That one cannot rightfully limit the amount of clothing and of covering. Chap. VI.

BOOK IX
CONCERNING APPETITE AND THIRST

BOOK X
CONCERNING MEALS AND DIGESTION

DIVISION OF THE THIRD PART INTO BOOKS AND CHAPTERS

BOOK XI
CONCERNING EATING AND FOOD

BOOK XII
CONCERNING PREPARATION AND ORDER IN SERVING FOOD

BOOK XIII
CONCERNING FRUITS, SALADS, AND CHEESE

How one should understand it when people say: *Post crudum purum.*[9] — Chap. VI.

That salad should be much stronger in salt than in vinegar; and why people say that it takes four people to make it properly. — Chap. VII.

Why people say: "He who drinks no wine after salad is in danger of becoming sick [*malade*]."[10] — Chap. VIII.

That lettuce is more healthful with honey than any other way. — Chap. IX.

Whether it is true that apples, pears, and walnuts ruin one's voice. — Chap. X.

Why people say that after an apple no man ever drank, and after a pear, drink or call a priest.[11] — Chap. XI.

Whether it is right to say that one must drink after cheese and after a pear. — Chap. XII.

Whether it is right to say that a pear with cheese is a marriage.[12] — Chap. XIII.

Why pears and apples are prized and valued without qualification. — Chap. XIV.

Whether it is right to say that the true shields against death are bread and cheese; nevertheless, people say that cheese is good when it comes from a stingy hand, and he who eats less cheese or ham puts one over on his companion. — Chap. XV.

BOOK XIV
CONCERNING DRINK

Whether it is good to eat a lot before drinking and (as they say) lay a good foundation. — Chap. I.

Why people say that drinking while eating one's soup ruins the teeth, and in Germany that it causes goiters to develop. — Chap. II.

Whether it is better to drink little and often during a meal, or large amounts. — Chap. III.

Whether it is a bad thing to drink before going to bed. — Chap. IV.

BOOK XV
TREATISE ON WINE

DIVISION OF THE FOURTH PART INTO BOOKS AND CHAPTERS

BOOK XVI
CONCERNING RETIRING AND SLEEPING

BOOK XVII
CONCERNING THE CAUSES OF DISEASE

BOOK XVIII
CONCERNING DISEASES

BOOK XIX
CONCERNING PROGNOSIS IN ILLNESSES

BOOK XX
CONCERNING FOOD DURING ILLNESSES

DIVISION OF THE FIFTH PART INTO BOOKS AND CHAPTERS

BOOK XXI
CONSIDERING THE CURING OF ILLNESSES

BOOK XXII
CONCERNING ABUSES AND REMEDIES

Whether washing the hair moistens more than it desiccates, unless it is dried in the sun. Chap. V.

Concerning those who keep recipes of remedies that have served them well all their lives and who share them with others. Chap. VI.

BOOK XXIII
CONCERNING POOR TREATMENT AND EXTRAVAGANT REMEDIES

Concerning the pernicious rule that one disorder cures another. Chap. I.

Against those who sow disorder in their illness in imitation of those who did not die from it. Chap. II.

Why people say that from one disorder come four orders. Chap. III.

Whether it is good to drink to satisfaction during an access of fever; and whether one must drink hot or cold liquids. Chap. IV.

Concerning those who drink a little pure wine on an empty stomach to ward off vertigo, migraine, and the shakes. Chap. V.

Concerning those who for a stomachache apply a cold tin plate. Chap. VI.

Concerning those who for colic apply a wet towel on the belly. Chap. VII.

BOOK XXIV
CONCERNING SUPERSTITIOUS AND USELESS REMEDIES

Against those who are wholly content with the efficacy of potions, without purgations or other remedies. Chap. I.

How it is possible to reset a dislocation without seeing or touching the patient. Chap. II.

Concerning conjured water, banners, lint candles, and conjured lard for the healing of wounds and ulcers. Chap. III.

BOOK XXV
CONCERNING GOOD AND TRUE REMEDIES

How one should understand it when people say: "For a headache, a stoppage of wine."	Chap. VIII.
Why people say that the mother's sickness requires the father.	Chap. IX.

DIVISION OF THE SIXTH PART INTO BOOKS AND CHAPTERS

BOOK XXVI
CONCERNING ORDINARY EVACUATIONS

Against those who get accustomed to vomiting every day.	Chap. I.
Against those who spoil their stomachs with mollifying things in order to have loose bowels.	Chap. II.
Concerning those who walk barefoot on a cold surface in order to have loose bowels.	Chap. III.
How one should understand the meaning of "having a good stomach."	Chap. IV.
Which is worse, being constipated or having loose bowels.	Chap. V.
Against those who are never quite comfortable except when their bowels move frequently.	Chap. VI.

BOOK XXVII
CONCERNING PURGATIVES OR DRUGS

Against those who, as an argument against drugs, cite the old age of people who never took any.	Chap. I.
Against those who refuse drugs as a precaution, saying that it is a bad habit to acquire.	Chap. II.
That purgation is indicated in every season, even during dog days.	Chap. III.
That children and pregnant women can be purged.	Chap. IV.
Concerning those who refuse to take drugs, even juleps, saying that they find them distasteful.	Chap. V.

BOOK XXVIII
REGIMEN OF THOSE WHO ARE PURGED

BOOK XXIX
CONCERNING BLOODLETTING

BOOK XXX
CONCERNING DEATH

APPENDIX B
LIMINAL POEMS

TO L. JOUBERT, MOST CELEBRATED ROYAL PHYSICIAN AND PROFESSOR OF MEDICINE OF THE UNIVERSITY OF MONTPELLIER, FROM ETIENNE MANIALD,[1] PHYSICIAN FROM BORDEAUX [1578 and 1579]

Medicine is a divine discovery that extends the hours
Of life, and that drives away all life's ills:
It is accustomed to bestowing restored lives to the
Terrified masses, and increasing postponements of death.
Noteworthy antiquity, closer to the gods, cherished it,
And their posterity was enthralled by a love of this art.
Greeks, Arabs, Romans, Gauls, Germans, and Spaniards
Make it illustrious and embellish it in many ways.
Medicine had come to the highest peak of praise, and
The art had just completed its growth.
But the languid carelessness of the age overturns all,
Or else decay makes everything deteriorate.

Thus, honorable medicine suffers from its own collapse,
And now the dignity of the salubrious art is perishing.
The vulgar masses and impostors pour out noxious filth,
And a hastened death carries off their victims.
The one who is willing and able to shore up such ruins
Is rare, and no remedy is offered for this sickness.
Joubert, like the sunshine, marked with glistening rays, rises;
He alone takes up such a great burden;
Exposing the ghosts of error and the pretense of physicians,
He repairs and restores the lost honor.
Take heart, Joubert! Medicine will rise again by the effort
You make. Go on blessing the art with your writings!

TO THE MOST BRILLIANT PHYSICIAN, L. JOUBERT, FROM ETIENNE MANIALD

[In Greek, followed by a Latin translation; 1578 and 1579]

[Translation of the Greek:]
There are among the Gauls three suns of healing;
They have the highest honor among physicians.
Very learned Fernel[2] is the first and most beautiful light.
Sylvius *epiones* [= gentle?][3] is its second glory.
And Joubert, healer in deeds and teacher of wisdom,
As a great physician, now has the third honor.

[Translation of the Latin:]
Gaul has produced three suns of the art of medicine,
The pride and great splendor in the line of healers.
Learned Fernel is reckoned to be the first glory.
Sylvius *epiones* [= gentle?] was its second fame.
And to Joubert, who now with a clever mind sheds light upon
The art's secrets, the third palm is given.

TO JOUBERT, ROYAL PHYSICIAN, AN EPIGRAM BY DOMINIQUE REVLIN,[4] PHYSICIAN FROM BORDEAUX

[1578 and 1579]

Error oftentimes utterly destroys grace, life, and mind.
He who uncovers it preserves them in driving it away.

Go, therefore, teaching what is right; dispel the errors.
Keeper of the name, you will also thus be keeper of life
And spirit. What better gift could you give to the masses?
And what thanks can be returned for such great gifts?

FOR JOUBERT'S MOST LEARNED BOOKS ON POPULAR ERRORS, FROM JEAN GUYON[5]

[1578 and 1579]

Avenger of errors, protector and author of integrity,
So far above the vulgar crowd, you for whom their very thoughts long.
Long ago amidst great applause you gave your *Paradoxes*.
Behold, once again you bring forth a paradoxical work,
Which, with darkness dispelled, may illuminate the goods of life,
And which noble Hippocrates would have written joyfully.
I could believe that the omen of your name was prophetic for you;
Your name Joubert, is taken from the clear sunshine,
For just as the golden radiance of the sun dissipates all those
Obstructing clouds throughout the aerial regions,
So also do you, the bright one, through the Apollonian art,
Constantly dispel the clouds of error that roam in space.

TO M. JOUBERT ON HIS POPULAR ERRORS, A SONNET

[1579]

Through the obscure night is light made more beautiful;
More beautiful is virtue because of the vile action;
Ugliness makes beauty appear more beautiful to our eyes;
And through falsehood is the glory of truth more whole.
If the heavy and thick folds of those watery masses
Have for a long time rendered obscure the day of heaven's eye,
When it finally frees itself from them it is more radiant,
And all the more fiercely flames throughout its course.
Thus, through so many errors that a blind people flees,
Your mind (their sun which makes their night into day)
Flames more sharply and makes its appearance more beautiful.
Error, a fecund Hydra, makes a swarm of errors.
You, however, are the fortunate Hercules who meets them,

Combating a thousand errors that a thousand ages have made.

PIERRE DE BRACH[6]

JOUBERT DU CHESNE,[7] TUTOR, LORD OF LISERABLE, DOCTOR OF MEDICINE, AND HIS FORMER PRECEPTOR, TO M. JOUBERT [1579]

The Father with the golden face, who spurs me so violently
With the points of His holy fury
To sing now on my lute that which for you,
My muse, your disciple, both gives and vows to you.
This same Delphian with a double crown
Wreathes your own head with wisdom and honor:
The one carries praise, the other offers to you
The flower of all the remedies Europe contains.
Oh, sole worthy recompense from a loving father!
Oh, sole worthy present, from such an admirable son,
To posterity through his learned labors,
From a Joubert who, in spite of the envy of stupid commoners,
Wanted to uncover again the errors
That they commit, jeopardizing our miserable lives.

FROM PIERRE CHAMBON DE GOTZ FROM AGEN[8] [1579]

If the glory of the powerful Theban is lasting
For having strangled the injurious serpent
That for each head torn off it made two more appear,
Rendering all the neighboring lands uninhabitable,
What will Joubert's honorable knowledge do,
Which cuts at the abuse of a more hideous monster,
A monster all blear-eyed from ignorance and error,
Which hates day after day the truthful thing?
Joubert has therefore done more, not content
With restoring health through Apollo's art
To man, suffering from many diseases,
But in order to surpass all of the most excellent,

With the arguments that are contained herein,
He wanted to free the people from their madness.

TO MONSIEUR JOUBERT, FOR HIS WORK ON THE POPULAR ERRORS

[1578 and 1579]

Divine mind, which goes about marrying
The most serious things to those of pleasure,
And which draws profit from the leisure
Of the least wearisome of your actions,
Who will not call your hours the most enjoyable,
Your days and your years? And moved by a desire
Always to learn, who will not run to consult you,
Second Oedipus, on the knottiest of problems?
The heavens, angered by our sins,
Kept these beautiful secrets hidden
In the darkness of time, which consumes all,
Without the noble and gentle wit of Joubert,
Which, from the knowledge of his learned flint,
Lights that fire hidden from our century.

SAL. CERTON OF CHÂTILLON[9]

FROM THE SAME TO THE SAME

[1578 and 1579]

The profit, the pleasure, and the discipline
That, together, fuse the teaching, the recreation, and the fierceness
Of your fire, your amusement, and your doctrine
In the mind, the heart, and in the imagination
Animate, satisfy, and fill with affection
The young, the people, and the wise, who tremble over it,
Because of the art, the laughter, and the rod with which he drives them
Through his words, his games, and his rigor.
In this book the young, the wise, and the people learn,
Take pleasure, and are also corrected,
Full of gain, of pleasure, and of improvements,
Which to their great profit, relief, and advantage,
In it give them, speak to them, and clearly demonstrate
The heavenly language of the most learned Joubert.

FROM THE SAME TO THE SAME, A MEASURED ODE

[1578 and 1579]

Joubert, whom Apollo over all others holds dear,
Joubert, whom the skies have above all others gifted
With a fine intellect and great knowledge,
Filled with honor forever returning to you,
Either with a voice full of wisdom, you come
To unfold your treasure before us by snatching
From death's door and away from sleep
The soul crying out over the waters of Charon,
Or, mixing together a wearisome artifice (in a much
Smaller amount) and your heavy-light subject matter,
You come to us to reveal many a secret,
Melding pleasure and profit.
Your leisurely spirit is ever seen as you always
Go exposing some hidden piece of wisdom
With which you can one day help yourself, and then
Bring to the patient the much-needed assistance,
Oh, ungrateful one that you are not, and never were,
For the exquisite talents the heavens infused in you.
And however ungrateful will be the age to come,
It will never silence your glorious honor.
It will call out your name, it will sing your praises
To the highest heavens. Books and time itself
Will be filled with the great and exquisite renown
Which you deservingly gain for yourself before death.
And I, your herald and precursor,
Shall go before it, showing the way,
Proclaiming your splendor, and honoring
The present and the future of your great glory.

CROWNING ME WITH PRAISE[10]

FOR L. JOUBERT,
MOST ILLUSTRIOUS ROYAL PHYSICIAN
OF THE KING OF FRANCE AND OF POLAND,
AND ROYAL PHYSICIAN
OF THE KING OF NAVARRE,
FROM S. MILLANGES,[11] ROYAL TYPOGRAPHER
[1578 and 1579]

Alcides is, by his merits, carried up into the sky,
For this great soul first scattered fierce monsters
for the benefit of the whole world.
You, instructed in the art of Apollo, chase away with art
The blinding darkness and the horrid monsters of error.
As he was worthy, let therefore the white-haired poets
Exalt Alcides in song; and let Apollo, father of the poets,
sing your praises.

APPENDIX C
THE PRINTER'S LETTERS TO THE READER

THE PRINTER TO THE READER
[1578]

When I undertook the promotion of printing books in this province of Guyenne, previously destitute of this great good, I had thought, and was resolute, about keeping, in my printing, to French orthography, used by almost all the learned authors who have given us their works written in this language. But, finally overpowered by the authority of some very learned people who have done me the honor of bringing or sending to me their beautiful writing to be printed, I was forced to yield to their wishes and print their books with the orthography they judged to be the best.

This is why the readers of my printings must not think me a Proteus and an unstable person if they see, in the books that I have printed, sometimes the old orthography, which I revere above all others, and sometimes a new one, which I do not like at all.[1] I must also be excused if, unaccustomed to the orthography that Monsieur Joubert wishes to be respected in his books, I have allowed several words to be printed in different ways and otherwise than the aforementioned Sire Joubert wanted, such as *sçience, concepuoir, conçeption,* and other similar words with a ç with a tail; because I wanted *sçauoir, conçoiuent,* and other words in which the *c* coming before *a* or *o* is pronounced like an *s,* to be spelled with the aforementioned ç. There could be some other similar mistakes which, once we are accustomed to this new orthography, we shall avoid more successfully in the other books, as well as when we reprint the present ones.[2]

[SIMON MILLANGES]

S. MILLANGES TO THE READER
[1579]

Because Monsieur Joubert, speaking in the last four books of this first part on conception, reproduction, childbirth, lying-in, and reckoning of virginity, was often forced, in discussing the errors committed in such acts, to use words and expressions that seem a little obscene, it would be good if only married people were to read the interesting information that is put there for them in these books. And the religious, both men and women, and all those who wish to live chastely without getting married ought to leave entirely the reading of these books to those who are married. As for others, who do not care to hear about the shameful parts, they can skip over the chapters and passages marked with this sign: *.[3] Nevertheless, those who wish to preserve their health will find interesting and precious information concerning the subject both in the Table of Contents[4] and in the first and the last three books, which we have printed again.

APPENDIX D
JOUBERT'S APOLOGETIC LETTER TO THE QUEEN OF NAVARRE

TO THE MOST AUGUST QUEEN OF NAVARRE, KING'S DAUGHTER, KING'S SISTER, AND KING'S SPOUSE

You have heard, madame, my attack upon a goodly number of the *Popular errors in the area of medicine and of health regimens.* This is the first part that I presented to you in all humility and reverence, so that you might serve as judge in the matter, if it should please you to do so. I would fear the evil tongues of the envious, mean, and malicious who will manage to find it indecent that I suggest such a subject to Your Majesty, when I am forced in a few passages to touch upon matters that are too physical, such as when treating conception, pregnancy, birth, lying-in, and, especially, the reckoning of virginity. But knowing that one can speak decently (as I do) of all natural functions no less than of all the parts of the human body, even the most secret and the most hidden (called shameful, which chaste eyes in no way fear seeing in public, during dissections), and remembering what Dion[1] recounts about the most virtuous wife of the Emperor Augustus, Princess Livia Romania (who spared the lives of several men, who were about to be put to death for being entirely naked when she happened upon them, by saying that they were in no way different from statues), I thought, armed with such reasons as solid defenses, that the venomous tongues and teeth could not harm me.

May it therefore please you, madame, to accept everything in the best light, and to forgive my audacity, founded and built upon your affability and studious nature, which make it possible for you to find good and decent what the less sensible scorn and despise because they also have no sincere curiosity in letters or in virtue except in appearance.

May God confirm the station of Your Majesty, overwhelming it with His graces and His holy blessings.

Your most humble and most affectionate servant,

Laurent Joubert

APPENDIX E
PRINTER'S DOCUMENTS OF PUBLISHING RIGHTS

By special grace and privilege of the king given in Poitiers on the thirtieth day of August 1577, M. Laurent Joubert, first regent doctor and chancellor in the Université de Médecine at Montpellier, is permitted to designate such printer and bookshop as he pleases for the printing of all his works and books with the prohibition and banning of all others, whatever their quality and condition, from printing, selling, or distributing them during the time and term of ten years after the first printing of each work and book. The whole under penalty of confiscation of the books, of arbitrary fine, and of all expenses, damages, and interests. As is further contained in the letters patent of the aforementioned privilege, signed HENRY, [and further down] verified and registered at the presiding seat of Agenois, on the seventh of November 1577.

The aforementioned M. Laurent Joubert has allowed, by a document signed in his hand, S. Millanges alone, royal printer, to print the first part of his work of the Erreurs populaires au fait de la medecine, *for the time and term of five years, starting from the last day of the printing.*

APPENDIX F
CHAPTER HEADINGS OF "LA SANTÉ DU PRINCE"

I. Exhortation to the prince to be mindful of his health and regulated in his actions, because of the duty of his charge, his reputation, and the grandeur of his house.
II. That the prince owes it to himself to set up a certain regimen modeled after nature in order to keep healthy.
III. What regimen a prince can observe in his daily actions.
IV. On the charge the physician has with respect to a prince, and how important the prevention of illness is.
V. On the vigilant care the physician owes the prince, and of the good that comes from it.
VI. Confirmation of the preceding through well-known examples.

NOTES

NOTES TO THE INTRODUCTION

1. Pierre-Joseph Amoreux, *Notice historique et bibliographique sur la vie et les ouvrages de Laurent Joubert, chancelier en l'Université de Médecine de Montpellier au XVIe siècle* (Montpellier, 1814); J. L. V. Broussonnet, *Notice sur Laurent Joubert, professeur et chancelier de l'Université de Médecine de Montpellier* (Montpellier, 1829); E. Wickersheimer, "Un brave homme et un bon livre: Laurent Joubert et les *Erreurs populaires au fait de la médecine et du régime de santé,*" in *La médecine et les médecins français à l'époque de la Renaissance* (Paris, 1906).

2. Natalie Zemon Davis, *Society and Culture in Early Modern France* (Stanford, California, 1975); Gregory de Rocher, trans., *Treatise on Laughter,* by Laurent Joubert, (University, Alabama, 1980); Robert Cottrell, *Sexuality/Textuality: A Study of the Fabric of Montaigne's 'Essais'* (Columbus, Ohio, 1981). See the Selected Bibliography at the end of the Introduction for references to articles by Dulieu, Longeon, and Gourg.

3. Several erroneous dates have been given for Joubert's death. Unfortunately, the Library of Congress has retained one of them (1583). This explains the discrepancy between the date it continues to give for Joubert's death and the date recorded in recent studies. See Amoreux, 115, n. 9; and Dulieu, 142.

4. I am indebted to Barbara C. Bowen for this insight. See her study, *Les caractéristiques essentielles de la farce française et leur survivance dans les années 1550–1620* (Urbana, 1964).

5. The apologetic letters of Bertrauan and Joubert are translated in the Preliminary Matter of the present translation.

6. See Appendix F for a translation of the chapter headings of this manual.

NOTES TO THE PRELIMINARY MATTER

1. See notes 1, 3, and 4 of the Dedicatory Letter to Marguerite de France.

2. See note 1 of the Dedicatory Letter to Marguerite de France.

3. In a marginal note Joubert gives the following abbreviated references to the *Digesta*: L. foediss. C. de adult. & stupre. & L. cum vir, C. codex. See note 10 in chapter II of Book Three.

4. In a marginal note appear two biblical references: Romans 1 and Genesis 38.

5. Two scriptural references are given in a marginal note: Romans 14 and 1 Corinthians 6 and 8.

NOTES TO THE DEDICATORY LETTER TO MARGUERITE DE FRANCE

1. In the first edition of the *Erreurs populaires*, Joubert dedicated his work to Marguerite de France, better known as Marguerite de Valois (1553–1615), daughter of Henry II and Catherine de' Medici. She was the wife of Henry of Navarre, future Henry IV of France, and came to be known as *La Reine Margot*. However, due to criticism concerning the appropriateness of dedicating to a princess a work treating explicit sexual matters, Joubert substituted in later editions the name of Guy du Faur de Pibrac (1529–84), a magistrate and poet whose stoically Christian *Quatrains*, published in 1574, were very popular and went through numerous editions. The scandal centered around the appearance of the word *vit*; Joubert denies he used this vulgar word for the penis, invoking a printer's error for his Latin word for man, *vir*.

2. It is ironic that Joubert makes this point of divulging medical information to laymen; Dulieu cites the *Erreurs populaires* as a major scandal in that it revealed in the vulgar tongue information previously kept secret by writing in Latin.

3. "The Marguerite of Marguerites" is a pun in French and was frequently used in Renaissance French. The expression can translate as "the pearl of pearls" or "the daisy of daisies"; but it can also mean the "pearl" of the three queen Marguerites of the time and refers to a collection of poems written by the "first" Marguerite, Marguerite d'Angoulême (1492–1549), sister of Francis I, *Les Marguerites de la Marguerite des princesses* (1547).

4. Joubert is referring to Marguerite de France (1523–74), daughter of Francis I and Claude de France. She married Emmanuel-Philibert, the duke of Savoy (1528–80).

5. The reference is to Francis I, king of France (1494–1547).

6. The judgment of Paris, cited frequently in sixteenth-century letters as the paradigm of thorny adjudication, was popularized by Jean Lemaire de Belges (1473–1525?) in his *Illustrations de Gaule et singularités de Troye*, published between 1510 and 1513.

NOTES TO THE FIRST BOOK OF POPULAR ERRORS

BOOK ONE, CHAPTER I

1. In this chapter Joubert situates medicine's place in sixteenth-century thought. The preeminence of theology is clearly granted, but medicine claims a privileged second place in the hierarchical scuffle. Joubert argues effectively against the power of life and death held by magistrates, discounting it in favor of the physician's art.

2. An echo of the prevalent Renaissance belief that all creation was for man in a man-centered universe.

3. Joubert keeps to his clothing imagery, which stands for man's acci-

dents or possessions, over which magistrates do indeed have power. Joubert's relative contempt for the object of their profession, and admiration of his own, figure in his argument for the preeminence of the art of the physician.

4. Joubert's mention of the archetype of physicians is significant. Aesculapius was the mythological god of medicine, the son of Apollo and Coronis. His sons Machaon and Podalirius, as Joubert indicates, were physicians in the Greek army. Chiron taught the young Aesculapius the art of healing, and when the boy grew up he was renowned not only for curing the sick but for resurrecting the dead. Jupiter struck Aesculapius dead when the latter was restoring Glaucus to life, for fear that all men would eventually seek immortality. Aesculapius's descendants were the priestly cast known as the *Asclepiadai*. Their knowledge of medicine was regarded as sacred and was transmitted from father to son.

5. Joubert's note: Pliny, *Natural History*, Book 29, chapter I.

6. The reference is to Cicero, *De officiis ad Marcum filium libri tres*, commonly translated as *On Moral Duties*.

7. Hippocrates (fifth and fourth century B.C.) was an itinerant physician. Much of our knowledge of him is based on legend. An important collection of medical works and treatises came down from his school, several of which are probably by Hippocrates. The importance of his work in furthering the art of medicine is undeniable, and a cult developed around him during the Middle Ages and early Renaissance.

8. Aulus Cornelius Celsus was a Roman medical writer of the first century A.D. His only complete extant work is the *De medicina* in eight books. This work inspired Joubert's *Erreurs populaires* and his *Pharmacopea*. The first two books of the *De medicina* treat diet, a major concern of Joubert's, and the fifth and sixth touch upon pharmaceutical preparations. Celsus's work was highly esteemed during the Renaissance (and is still valued in our own day for several reasons). Renaissance physicians still subscribed to the ancient notion that philosophers were a weak lot because they spent too much of their vital humors thinking, a debilitating activity. See, for example, Joubert's *Treatise on Laughter*, 131–33.

9. Diocles (of Carystus in Euboea), a celebrated Greek physician of the fourth century B.C., wrote several medical works of which but a few fragments have survived. Praxagoras was a celebrated fourth-century B.C. physician from Cos. He was a member of the Dogmatics, renowned for his knowledge of anatomy and physiology. Chrysippus (of Cnidus), a physician of the fourth century B.C., was a student of Eudoxus, the famed astronomer, geometer, physician, and legislator who had studied with Archytas and Plato. His works are no longer extant, but he is quoted by Galen. Herophilus was one of the most celebrated physicians of antiquity. He was a pupil of Praxagoras and practiced human and animal dissection. Only a few fragments remain of his medical and anatomical works. Erasistratus, a pupil of Chrysippus's, was a famed physician and anatomist. He had numerous followers and founded a school in Smyrna (Ionia) bearing his name. Both he and Herophilus are said to have dissected criminals alive. Joubert draws this information from Celsus's Prooemium to the first book of the *De medicina*, 8–9.

10. Galen lived during the second century A.D. Born in Asia Minor, he

became a doctor and ministered to the athletes at Pergamum for six years before going to Rome, where his fame caused jealousy among the Roman doctors. One year later he left Rome but was called back by Marcus Aurelius. His doctrine enjoyed a revival during the Renaissance; besides being famous for his theriaca, a much-used remedy composed of drugs and honey, he was often set in opposition to Aristotle by proponents of empirical medicine. He considered anatomy the basis of the art of healing.

11. Joubert's note: Exodus 15.

Joubert is referring to the episode in Exodus just after the crossing of the Red Sea, when the bitter water of Marah was sweetened by throwing a tree into it.

12. Joubert's note: Pliny, Book 26, chapter 2.

The reference of the passage from Hippocrates is *The Law* [*Nomos*], I–IV passim.

BOOK ONE, CHAPTER II

1. This dialectical method of argumentation is typical of Renaissance treatises in general and of Joubert's writings in particular. In his treatise on gunshot wounds (*Traitté des arcbusades*, 1570), for example, Joubert argues *pro* and *contra* the question of gunshot wounds as burns, the toxicity of gunpowder, and several other controversial subjects.

2. Job 14:5.

3. Joubert's note: Chapter 10.

4. Joubert's note: Chapter 10.

5. Averroës (Ibn Rushd) was a twelfth-century A.D. Spanish-Arabic philosopher and physician. The greatest of the Arabic Aristotelians, he greatly influenced Christian thinkers, and his work, although condemned both by the church and by the Muhammadan clergy, was highly esteemed by Renaissance physicians and philosophers.

6. Avicenna (980–1037) was an Iranian physician and philosopher. His works were part of the program for medicine in French universities until the middle of the sixteenth century.

7. Joubert's note: Fourth Book of Kings, chapter 20.

Joubert mistakenly writes "Elias" (*Elie*) for "Isaias" (*Esaïe*).

8. Herodicus, a fifth-century B.C. physician from Selymbria (Thrace), was one of Hippocrates's tutors.

9. Plutarch (of Chaeronea) was a first- and second-century A.D. philosopher, biographer, and moralist. He was particularly esteemed during the Renaissance, influencing writers such as Montaigne and political figures such as Henry IV.

10. Malmsey (Latin name *malmasia*) was a sweet, rich wine produced in the Mediterranean region from the malvasia grape. According to Cotgrave, imperiale was "the name of an hearbe (mistaken by some modern Herbarists for Smyrnium, or Candie Alexanders) that hath many excellent virtues." He defines alkermes as "a confection made of the decoction and infusion of silke into the juice of the grain *chermes*; a soveraigne remedie for all swond-

ings." "Swondings" is a spelling variant for "swoundings," (i.e., fainting spells). Cotgrave calls motherwort (*Leonurus cardiaca*) "good against the throbbing, or excessive beating of the heart."

11. Democritus, of the fifth century B.C., was known as "the laughing philosopher." He lived in Abdera, where the inhabitants, thinking him mad because he laughed ceaselessly, called upon Hippocrates to come and heal him. Hippocrates found him exceptionally wise when Democritus told him he was laughing over human folly and stupidity. The episode is recounted in Hippocrates's letter to Damagetes, which was translated from the Greek by Joubert's colleague at Montpellier, Jean Guichard, and included in the 1579 edition of the *Treatise on Laughter*. The Thesmophoria feasts were celebrated in Athens in the month of Pyanepsion. Thesmophoros (or Thesmia), that is, "the lawgiver," was a surname of Demeter (Ceres) and Persephone (Proserpina).

12. Joubert's note: After the seventh paradox of the first decade.

Joubert treats the question of God's will in his *Paradoxorum decas prima atque altera* (1566), a medical work in which various subjects are treated, such as hematosis, blood revulsion, innocuousness of menstrual blood, how each part of the body uses a particular humor for its sustenance, etc. See Dulieu, 145, and Amoreux, 64–76.

13. Sixteenth-century medicine condemned frequent copulation. Joubert cites as a cause of gout frequent or violent copulation while standing. Cf. infra, Book One, chapters III and IV.

BOOK ONE, CHAPTER III

1. A crown was worth more than four times the value of a teston.

2. Ulcers refer here to external sores. This sentence reflects both attitudes and practices of sixteenth-century medicine in France. The dichotomy between doctors and surgeons is noted. It fell to surgeons to do minor and major surgery; the doctors prescribed drugs and wrote treatises. The practice of bloodletting and maintaining lesions to purge the body of undesirable humors was common.

3. The term *empiric* or *empirical* was pejorative in Renaissance medicine. It referred to those who practiced the art of healing on the basis of experience rather than principles, and therefore in ignorance of the basis or consequences of any of their actions. See Joubert's *Treatise on Laughter*, translator's preface.

BOOK ONE, CHAPTER IV

1. Guy de Chauliac, a French physician of the fourteenth century from Avignon, wrote one of the standard works on surgery. This Latin work (*Chirurgia magna*) was translated into a "barbaric French" in 1478 and was known as the *Grande chirurgie*. It was improved upon in 1543 by Jean Tegault, a Paris physician. Numerous editions of the work appeared, including an English one in 1541, before Joubert's definitive annotated translation was pub-

lished in 1578 or 1579. See Amoreux, 54–59; Dulieu, 152; and Nicaise, passim.

2. Joubert, a Protestant, was not sympathetic to the Catholic views on the intercession of saints.

BOOK ONE, CHAPTER V

1. Quinsy was what is today called suppurative tonsillitis. On the matter of bloodletting, see note 2 of chapter III above.

2. On Herophilus, see note 9 of chapter I above.

3. Ecclesiasticus 38:2–3.

4. On Hippocrates's letter to Damagetes, see note 11 of chapter II above.

5. *Passato lo malo, poi è gabbato lo santo [sic]: "Once the illness is over, the saint is mocked."*

BOOK ONE, CHAPTER VI

1. Tertian agues (fevers) recur every third day. Celsus discusses them at length in the third book of the *De medicina* (3–17). Cotgrave defines as "continuall" a fever "whose fit neuer ceaseth till the disease, or diseased, end."

2. Light fevers.

3. According to the theory of humors, which followed closely the theory of the four elements, physiological conditions resulted from various proportions and concentrations of the four humors. See note 2 of chapter VI in Book Five below.

4. Joubert alternates between addressing laymen, physicians, and non-physicians both directly and indirectly. In this sentence he has switched to referring to a layman in the third person.

5. Certain remedies that would today be considered superstitious were in use during the sixteenth century, such as carrying amulets or reading the life of Saint Margaret to a woman about to deliver. Joubert debunks some of the least Christian ones but firmly believes in the power of the evil eye. See his *Treatise on Laughter*, preface to the second book.

6. Although the role of belief and the imagination seems to be discounted by Joubert, he elsewhere recounts convincing examples of the power of the soul over the body (*Treatise on Laughter*, preface to the second book).

7. Joubert is revealing part of the internal code of ethics and courtesy honored by physicians. It was doubtless one of the many reasons Joubert was criticized for publishing the *Erreurs populaires*.

BOOK ONE, CHAPTER VII

1. This chapter appears neither in the 1578 edition nor in editions following the 1579 edition. Possible reasons for its exclusion are given in the Introduction. See also note 8 of this chapter below.

2. Terence, *Adelphoe*, l. 98: "*homine imperito numquam quicquam iniustiust* . . ."

3. See note 1 of chapter VI above.

4. Joubert treats the question of gunshot wounds in his *Traitté des arcbusades*. See note 1 of chapter II above.

5. Joubert uses the same argument in his *Treatise on Laughter* when describing the infinite number of types of laughter (pp. 87–90). In the three sentences that follow, Joubert draws upon Celsus's Prooemium to the first book of the *De medicina*, 52–53.

6. The notion of difference, not only between oneself and others but also between oneself at different moments, fascinated Joubert. It was to be developed by Montaigne in his *Essais* (II, i). For Joubert's influence on Montaigne, see Cottrell.

7. The deficiency in weight of a half-grain apparently compromised the value of an *écu*, a crown of sterling.

8. Criticism of this sentence may explain in part the absence of this chapter in future editions of the *Erreurs populaires* (those of 1584 and 1587, for example). It is not certain whether this chapter was withheld from the 1578 edition or simply had not yet been written. The former hypothesis seems more likely because virtually all of Joubert's works were composed several years before they were published.

BOOK ONE, CHAPTER VIII

1. Like the preceding chapter VII, this chapter did not appear in the 1578 edition or in editions following that of 1579. Reasons for its exclusion may spring not only from negative comments about physicians, but also from indirect mention of his involvement in the illness of a marshal of France, possibly Artus de Cossé (1512–82), made marshal of France in 1567. L'Estoile says that he was called the Marshal of Bottles because of his love of food and festivities.

2. The notion of inconstancy, a frequent theme of the time, will play an essential role in Montaigne's *Essais*, especially in the "Apologie de Raymond Sebond" (Book II, chapter xii). See Cottrell.

3. Another sentence Joubert may have later regretted. See note 1 of this chapter and notes 1 and 8 of chapter VII above.

4. Joubert manages to lay a large portion of the blame for abuses in medicine at the feet of the magistracy, the other professional group Joubert saw as vying for power in sixteenth-century France.

5. Celsus, *De medicina*, Prooemium to the First book, 46–50.

6. See note 1 of this chapter above.

7. Cotgrave defines the purples (*le pourpre*) as "a pestilent Ague which raises on the bodie certaine red, or purple spots; also the Tokens; the blew spots appearing on a bodie thats mortally infected with, or dead of, the Plague."

8. This paragraph must have elicited criticism; the entire chapter, we recall, is absent from future editions.

9. Celsus, *De medicina*, V, 26, 1C–D.

10. One of several instances of statements at cross-purposes with Joubert's interests, and therefore another possible reason for this chapter's exclusion from subsequent editions.

BOOK ONE, CHAPTER IX

1. Joubert's use of the siege of Sancerre in 1573 as a metaphor for stubborn resistance would have struck the imagination of sixteenth-century readers. Also, since he was a Protestant, it could have given him pleasure to recall this heroic moment of the Reformation. Earlier in this same year the siege of La Rochelle had taken place, leading to the signing in June of the Treaty of La Rochelle.

2. Setting afire besieged cities was a common tactic used to bring about surrender, but the reference may be to a specific rumor. Historian Jules Michelet reminds us that at the end of 1575 it was believed that Montmorency-Damville was coming to "burn everything within twenty leagues of Paris" (*Histoire de France*, 3 vols., Paris, n.d., II, 1509).

3. Here and in the fourth chapter Joubert walks a fine line in telling readers, on the one hand, to accept divine will and, on the other, to employ every possibility at their disposal to fight death. The notion of "tempting God" is closely related to the refusal to observe and use nature's laws and cures. Nature, as Joubert notes often, is informed by God.

4. The allusion is to Matthew 4:1–11.

BOOK ONE, CHAPTER X

1. It is clear in Joubert's French that he is momentarily adopting the layman's manner of reasoning.

2. Again, Joubert argues in laymen's terms against one of the informal fallacies: *Post hoc ergo propter hoc.*

3. A word-for-word rendering of what would be the more idiomatic English: "That's a horse of a different color."

BOOK ONE, CHAPTER XI

1. Although Alfonso II d'Este (1559–97), the last duke of Ferrara, was roughly Joubert's contemporary, it is Alfonso I d'Este (1505–34) who is referred to here. Gonnella, mentioned a few lines later, was a stock buffoon in Italian literature, appearing, for example, in Poggio Bracciolini's *Liber facetiarum*, usually referred to as the *Facetiae*. I am grateful to Barbara C. Bowen for pointing out to me that although written in the 1450s, this collection was not published until 1470 in Florence (see her article, "Renaissance Collections of *Facetiae*, 1344–1490: A New Listing," *Renaissance Quarterly* 39:1, 1–15). The story is from item 165 in Poggio: "Facetissimum histrionis Gonellae." There were also historical figures with the name of Gonnella who were at various times in the service of the Este household.

2. Antonio Musa Brassavolo (1500–1555) is cited here as the learned physician par excellence. He wrote, among several other books, the widely read *Examen omnium simplicium medicamentorum*, a pharmaceutical work published in Rome in 1536. After serving as physician to Alfonso I, he went to France.

3. Terence, *Adelphoe*, 99: *Qui nisi quod ipse fecit nil rectum putat*. See note 2 of chapter VII above.

BOOK ONE, CHAPTER XII

1. As in previous cases, I have kept to the French saying literally. A more idiomatic English would be: "Two heads are better than one."

2. The saying is from Hesiod's *Works and Days* [*Erga kai hemerai*] (l. 25), a work that, along with the *Theogony*, Joubert read and from which he occasionally cited passages. Cf. his *Treatise on Laughter*, 14, 88, 89, and 135.

3. Celsus, *De medicina*, III, 4.6–10.

BOOK ONE, CHAPTER XIII

1. On the letter of Hippocrates to Damagetes, see note 11 of chapter II above.

2. Joubert is obviously uncomfortable arguing both against the behavior of patients who bid higher and higher for the services of the popular doctor and against the questionable ethics of the physician who charges more due to heavy demand. Curiously, Joubert does not entertain the possibility of simply having fewer patients. The reason is unclear, because he usually foresees such objections by virtue of his *pro* and *contra* discursive tactics and then disposes of them.

3. The celebrated physicians who overcharge patients are attacked only obliquely by Joubert, since his solution is to enjoin patients to engage less renowned physicians.

BOOK ONE, CHAPTER XIV

1. See note 6 of chapter VI above and Joubert's *Treatise on Laughter*, 65–72.

2. This is a notion alluded to by Rabelais in his letter to Monseigneur Odet (made cardinal of Châtillon in 1553) and in the *ancien prologue* to the *Quart livre*. See Rabelais (1), II, 3–9, and II, 569–78.

3. I beg the reader's indulgence here and in other instances where I have attempted to keep the ternary and parallel structures of this sentence, typical of Joubert's language and of that of sixteenth-century prose in general.

4. Joubert's emphasis on the power of belief in curing illness is a necessary stay in his argument: the chapter evolves from a profound bow to the premier physicians at the opening to a plea at the end for putting the same therapeutic trust in physicians of lesser fame and success.

BOOK ONE, CHAPTER XV

1. Developed by Joubert a few sentences later.

2. Joubert's note: Ecclesiasticus 38[:4].

3. Joubert's use of the term *rhabarbatiuement* was still highly metaphorical in the sixteenth century; it would translate literally as "rhubarbly." Modern French has the adjective *rebarbatif*, meaning gruff, but the standard etymology given it (*se rebarber*) would seem to be in error.

4. This passage from Celsus and the two sentences following it are missing in the 1578, 1584, and 1587 editions. The exclusion might be attributable to concern over the questioning of the physician's authority. Joubert finds himself undermining the foundation of his own profession, the prescription of drugs. Any attack on the safety, value, or necessity of pharmaceutical preparations would be a threat to the physicians' power.

5. We recall Joubert's constant qualification of the healing or saving of lives by physicians as done through the grace of God. He is ever vigilant not to stake out for medicine an area that the contemporary theological order occupied.

6. The proverb to which Joubert is referring is no longer widely used. The Phrygians, along with the Boeotians, were known in antiquity for their stupidity. Cicero speaks of them (*Pro L. Flacco oratio*, 65) as becoming better if they are beaten: "*Phrygem plagis fieri solere meliorem.*" Sextus Pompeius Festus, a Roman grammarian of the second (or possibly the fourth) century A.D., wrote a compendium of aphorisms, which was republished by Scaliger (a humanist and a contemporary of Joubert's), and according to the *Oxford Latin Dictionary*, he is the author of the proverb taxing the Phrygians with the flaw of learning too late: "*. . . 'sero sapiunt Phryges' proverbium est natum a Troianis, qui decimo denique anno velle coeperant Helenam . . . reddere Achivis.*" Barbara C. Bowen's point is well taken: many of the proverbs from antiquity that Renaissance authors cited were via Erasmus's *Adages*.

BOOK ONE, CHAPTER XVI

1. This entire chapter is missing in the 1578, 1584, and 1587 editions. Possible explanations are given in the following notes and in the Introduction.

2. This portrait of women is characteristic of the misogynistic tradition prevalent in the sixteenth century. Joubert's anger stems from their blatant disobedience of physicians' orders concerning patients' food.

3. This passage, although hypothetical, may explain in part its exclusion from subsequent editions.

4. This passage from Celsus (*De medicina*, III, 4.8–10) was already cited by Joubert in chapter XII above. Joubert omits from the citation a passage recalling what Erasistratus said concerning the close surveillance of the patient.

5. See note 3 of this chapter.

6. Joubert's tone in French is highly ironical. See note 4 of chapter I above.

7. The implication that women's stubbornness and ignorance had fatal results obviously did not go uncontested. The matrons' opposition to the *Erreurs populaires* was strong; see the opening paragraphs of chapter IV of Book Five. Joubert excluded this chapter from later editions.

8. These are not distillates as we know them but, rather, broths. Joubert describes the process a few paragraphs later.

BOOK ONE, CHAPTER XVII

1. This attitude is the foundation of sixteenth-century medicine. It is expressed more clearly by Joubert at the opening of his *Traitté des arcbusades* (p. 1) in a citation from Galen: "One cannot begin to determine the most elementary curative application for any disorder whatsoever without first ascertaining its exact essence." The passage is found in the *Peri physikon dynameon* (*De naturalibus facultatibus*), II, ix.

2. Joubert's explanation of the function of anesthetic agents, cast in terms of the theory of the four humors, necessarily involves moisture and natural heat.

3. Chapter X above.

BOOK ONE, CHAPTER XVIII

1. 1 Timothy 1:9.

2. Joubert's note: Matth[ew] 9[:12].

3. Joubert omits the "rubber and anointer" Celsus mentions here: "... *neque medico neque iatroalipta* ..."

4. Joubert's note: Book One, chapter [section] 1.

The following two passages from Celsus are those omitted by Joubert in his citation: "... But whilst exercise and food of this sort are necessaries, those of the athletes are redundant; for in the one class any break in the routine of exercise, owing to necessities of civil life, affects the body injuriously, and in the other, bodies thus fed up in their fashion age very quickly and become infirm ..."; "... Since, however, nature and not number should be the standard of frequency, regard being had to age and constitution, concubitus can be recognized as not inopportune when followed neither by languor nor by pain. The use is worse in the daytime and safer by night; but care should be taken that by day it be not immediately followed by a meal, and at night not immediately followed by work and watching."

BOOK ONE, CHAPTER XIX

1. This paragraph, obviously designed to whet the reader's appetite, also reveals laterally both Joubert's concern over the negative effects certain portions of his work might have in the medical order, and the painful deliberations leading up to the exclusion of passages and chapters in the revised edition.

2. Joubert is referring to the harm done to the sick not only because of the ignorance of laymen, but because of that of physicians.

3. It appears that Joubert is mounting a facile attack upon women, but such is not the case; women commonly filled the ranks of attendants in sixteenth-century France, the particular group singled out here. This is not to deny the presence of misogyny in Joubert or during the Renaissance, but to show another aspect of the basis for his frequent attacks on women. Toward the end of the chapter, he also attacks apothecaries, among whose ranks no women were to appear until 1869.

4. See note 3 of chapter III above. Cotgrave defines the empiric ("Empiricke") as "a Physition which without regard either of the cause of a disease, or of the constitution of the patient, applies those medicines whereof he hath had experience in others, worke they how they will."

5. The information in this paragraph is most revealing with respect to medical practices in France, Italy, and Spain. For a discussion of practices in these countries, as well as in Germany and Portugal, see Brabant, 189–207 and 241–71.

6. The authoritarian tone is characteristic of physicians in sixteenth-century France. For a discussion of the economy and display of power by Renaissance physicians, see the Introduction.

BOOK ONE, CHAPTER XX

1. Joubert uses the word *écu*, which I have translated as "crown" throughout, except in a few passages where it is used in opposition to *livre* (in which case I have kept *écu*). An *écu* was usually worth three *livres*. Since the proportions are the same between them, I have used the English terms "pound," "shilling," and "penny," respectively, to translate *livre*, *sol* (*sou*), and *denier*: twelve *deniers*/pence = one *sou*/shilling; twenty *sous*/shillings = one *livre*/pound. See note 1 of chapter III and note 7 of chapter VII above and note 5 of this chapter.

2. While Joubert appears to be vaunting the apothecary's trade, it is the physician who authorizes the transformation of the apothecary's product from raw material to valuable commodity. The tone of the chapter soon changes decidedly as Joubert mounts an acerbic attack on apothecaries for encroaching upon what Joubert stakes out as the physician's domain.

3. Joubert's note: Popular Error I.

These marginal notes enumerate the fifteen errors announced in the chapter title as they come up in the discussion.

4. Joubert's note: Popular Error II.

5. As stated above (note 1 of this chapter), I am translating Joubert's "*liures . . . sols . . . deniers*" as "pounds . . . shillings . . . pence." However, their worth relative to English currency must be decreased by a factor of ten, as Cotgrave notes: (1) "Livre Tournois. . . . the most ordinarie French pound; amounts but to 2 s. sterl." Since there were twenty shillings to the pound, its worth was one tenth that of the English pound sterling.(2) "Sol. . . . A Sous, or the French shilling; whereof tenne make one of ours; But this is to

be understood of the *Sol Tournois*, the most generall, and best-known *Sol* in France; and euer understood when the word *Sol* is used without addition."(3) "Denier. . . . A pennie; a deneer; a small copper coyne valued at the tenth part of an English pennie . . . "

6. Joubert's note: Popular Error III.

7. Joubert's note: Popular Error IV.

As a rule, all medical works were written in Latin. Translation into the vulgar tongue began in earnest during the Renaissance, and some, like Joubert's *Erreurs populaires*, generated much interest as well as controversy. One might draw the analogy between this vulgarization and that done by Blaise Pascal in the next century, when hitherto obscure theological questions were cast in comprehensible language in the *Provinciales*.

8. Joubert's note: Error V.

9. See note 9 of chapter I above.

10. *Usure*, translated as "usury," was a common term during the Renaissance, designating use or interest, and is not pejorative here—a point which Joubert's parenthetical insert could serve to demonstrate.

11. Joubert gives us a considerable amount of sociocultural information in this chapter. It is all the more precious in that this chapter does not appear in any other edition of the *Erreurs populaires*. Joubert's vigorous defense of physicians' rights soon becomes apparent in this supposed apology for the apothecary. Indeed, with the very next paragraph the tone changes, and Joubert criticizes apothecaries for amassing huge clienteles and for attempting to diagnose and treat patients empirically.

12. Joubert's note: Error VI.

13. Joubert's note: Error VII.

14. Cotgrave explains this proverb: "Chanter Magnificat à matines. *To doe things disorderly, or use a thing unseasonably.*"

15. See note 3 of chapter III and note 4 of chapter XIX above.

16. Joubert's note: Error VIII.

Saladino da Ascoli was a fifteenth-century Italian physician in the service of the prince of Taranto (Giovanni Antonio del Balzo Orsino) and author of the *Compendium aromatariorum* (Bologna, 1488), the classical pharmaceutical work for nearly a century during the Renaissance, until Joubert's *Pharmacopea* appeared, first in Latin (Lyons, 1579), and shortly thereafter in French (Lyons, 1581). Saladino must not be confused with Saladin (as Joubert writes), or Salah-ad-Din, the twelfth-century sultan of Egypt and Syria. There might be some room for error in that Saladin's celebrated physician, Maimonides (Moses ben Maimon), wrote a twelfth-century treatise on poisons, which was known during the Renaissance. See Brabant, 100–101.

17. Nicolas Prevost (Nicolaus Prepositi) was a French physician and author of at least two pharmaceutical works, the *Dispensarium . . . ad aromatarios nuper diligentissime recognitum* (Lyons, 1528) and the *Isagoge, sive introductiones in artem apotecariatus incipiunt* (n.p.,n.d.).

18. Joubert's note: Error IX.

19. Joubert's note: Error X.

20. Joubert's note: Error XI.

21. Joubert's note: Error XII.
22. Joubert's note: Error XIII.

Error XIV appears on the next page, incorrectly paginated 148 for 147 in the 1579 edition.

23. In chapters XI and XIX above.
24. Joubert's note: Error XIV.
25. I have kept the term *écu* here instead of the standard "crown" to establish the difference between what the physician earns as opposed to the apothecary. The physician earned three times more than the apothecary.
26. Joubert's note: Error XV.
27. Chapter V above.
28. The ratio here is 5 or 6 to 1.
29. Assuming the painter also took eight days to complete the portrait, the ratio is 18.75 to 22.5 to 1—a considerable difference.
30. As pointed out in note 3 of chapter XIX above, attendants were nearly always women.

NOTES TO THE SECOND BOOK OF POPULAR ERRORS

BOOK TWO CHAPTER I

1. This conclusion proves to be temporary and partial.
2. Joubert follows Aristotle's notion of the father's sperm producing the child's essence, while the mother supplies the matter in the form of menstrual blood (*De generatione animalium*, I, xvii–xx). Galenists, however, maintained that women also produced sperm (on the Renaissance debate over the presence and efficacy of female sperm, see Maclean, 28–46). Aristotle's theory of conception was maintained through the Renaissance (see, for example, Jacob Rüff's *De conceptu et generatione hominis* [Zurich, 1554]) and was first seriously questioned in William Harvey's *Exercitationes de generatione animalium* (London, 1651 [another edition appeared in the same year in Amsterdam]). But it was only with the use of the microscope in the second half of the seventeenth century by Antoni van Leeuwenhoek (1632–1723), who published his discovery of spermatozoa in 1677, that it was possible to ground such theory in observation (for details of Harvey's postulation, see Eccles, 40–41).
3. Joubert's explanation of the menses is based on the theory of humors. See note 3 of chapter VI in Book One.
4. Joubert thus dismisses a popular belief that holds that a nursing mother rarely conceives, giving as explanation the presence of sufficient blood in the womb to nourish the sperm.
5. Presumably, so as not to rob the recently conceived child of its nourishment.
6. Madame de Montluc was the wife of Blaise de Lasseran Massencôme de Montluc (or Monluc) (1502–77), a captain whose battle experience was

extensive and who was made marshal of France in 1574. After a severe face injury he turned to writing his memoirs, the *Commentaires*, published posthumously in 1592.

7. Joubert discusses them in later chapters.

8. Joubert expresses here a notion that Montaigne will develop in discussing savages and the constraints of civilized societies in "Des cannibales" (*Essais*, I, xxxi). See note 2 of chapter VIII in Book One.

BOOK TWO, CHAPTER II

1. Ferrante was the fourth prince of Salerno. He followed Charles V to North Africa and to Spain, was captain of the Italian infantry in Lombardy, and served as Neapolitan ambassador to Spain before being invited to the court of Henry II of France. Considered a traitor by the Spaniards, all his land was confiscated, and he died in Avignon dispossessed and without heir.

2. Until 1562, the king of Navarre was Antoine de Bourbon (1518–62), in which year his son, Henry III and future Henry IV of France, became king of Navarre.

3. Joubert differs from what Aristotle says in the *De generatione animalium* (I, xix) on the matter of puberty in males and females. Without specifying age, Aristotle maintains that men produce semen and women menstrual discharge "at the same time of life." In the *De politica* (VII, xv) Aristotle seems to say that puberty occurs at fourteen years of age.

4. Joubert is mistaken as to the young poet's nationality. Michele Verino, the precocious son of Ugolino Verino, was an Italian youth who died in 1487 immediately after composing his *Disticha moralia*, a conceit-ridden series of poems that gained him considerable notoriety.

5. Joubert's note: Aristotle, in his *Politics*, Book 7, chapter 16.

Since Book VII of the *De politica* has only fifteen chapters, Joubert is mistaken, meaning to put chapter 14, upon which this passage draws heavily. The 1587 edition also gives the following amusing marginal note: *Quand la fille pese un auque, on luy pot mettre la cauque (dit le vulgaire).* This would translate as: "As soon as a girl weighs as much as a goose she can be stuffed with a tent (according to the common expression)." Since a tent was a roll of linen used to dilate a canal or keep a wound open, the word refers here to the penis.

6. Joubert's note: Part 2, Book One, Epistle 63.

7. This last paragraph does not appear in the 1578 or in the 1584 edition but is in the 1587 edition.

BOOK TWO, CHAPTER III

1. See note 12 of chapter II in Book One above.

2. *De generatione animalium*, II, iv.

3. See note 6 of chapter I in Book Two above.

4. Although the *denier*'s size and weight had been fixed by Charlemagne at what today would be 21 millimeters and 1.7 grams, its use began to wane.

It decreased in size and weight until the seventeenth century, at which time it no longer served as a denomination. Thus, the diameter of the *denier* to which Joubert refers would have been approximately three-quarters of an inch.

5. Morphew was "a leprous or scurfy eruption. *Black, white morphew*" (*Oxford English Dictionary*).

BOOK TWO, CHAPTER IV

1. The subject of this generalization is the double proposition of the chapter title.

2. Joubert's note: Book 2 of the *Generation of Animals*, ch. 3, and Book 4, chap. 6.

These references are correct.

3. Joubert's note: See what Galen says of it in the last chapter of the first book of his *Faculty of Nutrition*.

Joubert must be referring to *On the Natural Faculties*, I, vi.

4. Joubert's comparison is difficult to understand today. It was currently believed that seeds could change to other strains or even other species under the influence of moisture. Ray and darnel are weeds that commonly infested wheat and barley fields. Way bennet is "Wild Barley-grass (Hordeum murinum)" (*Oxford English Dictionary*).

5. See note 2 of chapter I in Book Two above.

6. This sentence shows Joubert hedging on Aristotle's theory of conception, prevalent in the Renaissance but more seriously contested by Galenists toward the end of the century. Women can at times produce a better (male) offspring when a "phlegmatic" male's sperm would have normally produced female offspring.

BOOK TWO, CHAPTER V

1. Joubert's note: Chap. 15[:24].

2. For a discussion of leprosy during the Renaissance, see Brabant, 49–67.

3. Although Joubert speaks here of "sperm" coming from the woman, he states elsewhere that women provide no form or essence, but only the matter of the embryo. See notes 5 and 6 of the preceding chapter.

4. Joubert wrote *cordier* for *cordelier*, which I have translated as "monk." The *cordeliers* were members of the religious order of Saint Francis and were thus called because of the three-knotted cord with which they girded their habits. The story Joubert refers to is from Marguerite de Navarre's *Heptameron*, a collection of tales modeled after Boccaccio's *Decameron*. In this story, the monk gives a devout gentleman a "special dispensation" to sleep with his wife, who had had a child three weeks before, on the condition that he keep a separate bedchamber until two hours after midnight, supposedly so as not to trouble her digestion. Needless to say, the monk joins the young wife as soon as the members of the household retire for the night. This is

the part of the tale, far more complex in several other aspects, that interests Joubert.

BOOK TWO, CHAPTER VI

1. The proverb is from Terence's *Eunuchus*, l. 732 (*"sine Cerere et Libero friget Venus"*).
2. See note 3 of the preceding chapter.
3. The reference is to King David who, after committing adultery with Bathsheba, acknowledged his sinfulness (2 Samuel 12).

BOOK TWO, CHAPTER VII

1. Joubert uses the word *passereaus*, which literally means "sparrow" but which I have translated as "lecher" according to the sense, as Cotgrave will bear me out: "Passereaux, & moineaux sont de faux oiseaux. *Cocke Sparrows and (young) Monks are (much of a disposition) shrewd lechers."*
2. Joubert uses the word *espris*, which I have translated as "spirits," but I could also have used "humors." The distinction between the terms is minor, and they often appear as stylistic variants; spirits are humors that are finer and in motion. See Joubert's *Treatise on Laughter*, Introduction, xiii.
3. Joubert's note: Psal. 127.
4. Celsus, *De medicina*, I, i.

BOOK TWO, CHAPTER VIII

1. Joubert's note [1579 edition only]: Objection.
2. Joubert's note [1579 edition only]: Response.

Vapeur often translates as "breath," "smell," " aroma," "or odor" (Cotgrave gives "vapor, fume, exhalation, hot breath, steaming, reeking"). Since, however, it is a more active substance in this context, I have used "vapor."

3. Joubert's note [1579 edition only]: Objection. Response.
4. This same admiration characterizes Gargantua's famous letter to Pantagruel in Rabelais's *Pantagruel* (viii).

BOOK TWO, CHAPTER IX

1. See note 6 of chapter IV in Book Two above.
2. The French term for these plants is *herbes de la Saint Iean*. Joubert seems to treat them as inefficacious, if not the object of superstition, in the present chapter, but his attitude is not so negative when he notes the ancillary role of such remedies elsewhere, as in his *Pharmacopea* (Latin edition, 1579; French edition, 1581) or in his *Treatise on Laughter*, 15.
3. Joubert is being facetious. The mention of horns is a reference to cuckoldry, the category of humor that Joubert finds the funniest (*Treatise on Laughter*, 24).

4. The months without *r* in French are the same as in English: May, June, July, and August.

BOOK TWO, CHAPTER X

1. I have not adopted Natalie Zemon Davis's translation of *un quarton de son* as "a sheaf of straw" (p. 261) because *quarton* does not mean "sheaf," but rather "sheave." This is confirmed by Bernardin de Saint-Pierre, for example: " . . . *carton de son herbier* . . . " (*Harmonies de la nature*, 1814, p. 132 [cited in the *Trésor de la langue française*]). Although the term is used to designate various weights and measures, as is also the term *septier*, it cannot be made the equivalent of "sheaf" because Joubert establishes in his first paragraph a difference between a quarter of a *septier* of straw and a sheaf (*gerbe*, or the Gascon *glech*) of straw. Since the most that a *septier* of wheat could weigh was fifty-five pounds, the same volume of straw would weigh much less; indeed, the context argues for a much lighter weight. One is tempted to read *quarteron* as a quarter of a pound, or even a measure of wheat in Languedoc (Cotgrave), but all the editions insistently reproduce *quarton*. Most probably, however, it should be read as *quartault* (Cotgrave: "the Quarter, or fourth part, of a measure"), having undergone a nasalization rather typical in Joubert's dialect.

2. Joubert's note: Book 7, ch. 16.

3. Joubert's note: In St. Luke, chap. 1.

4. Joubert's note: Chap. 1.

5. Joubert writes "three leagues" (*à trois lieuës de Carcassonne*). A *lieuë* was, according to Cotgrave, normally two miles, but in the region of Languedoc it was three miles. Thus, Saint-Denis was about nine miles from Carcassonne.

6. This paragraph exists only in the 1579 edition.

7. That is, many who are the children of old men do not have deep-set eyes.

8. Joubert's silence at this point is not merely a matter of coyness or a contradiction of his usual practice of promulgating medical knowledge. It is a delicate question of public decency in confronting a strong taboo. The "hidden" reason for sunken eyes to which Joubert alludes is masturbation, condemned not only by Galen but by the ancient philosophers and physicians as well, on the grounds of expenditure of natural moisture. See Michel Foucault, *The Use of Pleasure, Volume 2 of The History of Sexuality*, trans. Robert Hurley (New York, 1985), 14–24 and 125–39, and the as yet untranslated third part, *Le souci de soi* (Paris, 1985), 121–69.

BOOK TWO, CHAPTER XI

1. I have kept the French designation of these herbs because of the reference to the saint's day (Saint John's). For the English names of these herbs, see note 2 of chapter IX in Book Two above.

2. The portion of this sentence from " . . . being so chafed . . . " to the end does not appear in the 1579 edition. It was probably removed in response to the violent reaction against the 1578 edition and its dedication to the Princess Marguerite de France. It was reinserted in later editions, in which the Dedicatory Letter was addressed to Pibrac. See note 1 of the Dedicatory Letter to Marguerite de France above.

3. See note 6 of chapter IV in Book Two above.

4. There is an unfortunate misprint in the 1579 edition, corrected in all other versions: *sinon que le mary fut mieus* [*vieus*]:"unless the husband is better [old]." It could be wit on the part of Joubert or the printer, Simon Millanges, but this seems highly unlikely.

5. Joubert used the same tactic in his *Treatise on Laughter* (p. 87) to argue for the infinite variety of types of laughter.

6. Joubert is possibly referring to the century plant *(Agave americana)*, which flowers once before dying after approximately fifteen years (mythically, after one hundred years).

7. Animal reproduction is discussed in the first chapter of Book Three.

BOOK TWO, CHAPTER XII

1. Although Cotgrave gives "a nephew; the sonne of a brother, or sister" for *neueu*, it is clear from the context that Joubert uses *neueus & [a] riere-neueus* in the classical sense (Latin *nepotem*) for grandchildren and great-grandchildren, and I have translated accordingly.

2. I have kept the French for its colorfulness. Cotgrave gives the following translation for the expression: "Le mortier sent tousiours les aulx: Pro[verb]. *Th'ill impressions made by nature, or bad habit got by custome, are seldome, or neuer, worne out.*"

3. I have translated Joubert's *tigne* (also spelled *teigne*) as "strangletare" (*Vicia hirsuta*). Other possibilities are "Tenentare, Tine, Strangleweed, Choakweed, Choakfitch" (from Cotgrave) and "tine-grass, tine-tare, and tine-weed" (*Oxford English Dictionary*).

4. The 1578, 1579, and 1584 editions bear the marginal notes 1 and 2, respectively, next to the passages in which discussion of each of the two questions begins.

5. Joubert never finished his treatise on syphilis, which was to have borne the title *De variola magna, siue crassa, gallis dicta* . . . His students gathered his lecture notes on the first part of his treatise, and a physician of Montpellier named Marc de Lacroix published them in Valence in 1582. See Amoreux, 98–99, and Dulieu, 156 and 162.

6. See Appendix A for the proposed structure of future parts and books of the *Erreurs populaires*.

7. "Not at twenty-four carats" is the literal translation of *nompas à vint & quatre quaras*. We would express such portions or probabilities in terms of percentages (twenty-four carats would be the equivalent of one hundred percent).

1. This chapter appears only in the 1579 edition. It may have been withdrawn because of certain passages that appeared graphic, if not scabrous, even for sixteenth-century readers. However, Joubert sought to divert his readers and often lightened the tone of his treatises with anecdotes. The reaction to those in this chapter must have been strong, especially in the light of the book's dedication to Princess Marguerite de France. See note 2 of chapter XI in Book Two and note 1 of the Dedicatory Letter to Marguerite de France above.

2. On Avicenna, see note 6 of chapter II in Book One above.

3. Among the several men bearing the name of Rufus, Joubert is most likely referring to Rufus of Ephesus (first century A.D.), whose fragmentary writings are preserved in part by Paul of Aegina, a seventh-century A.D. Byzantine physician known as the last of the Greek compilers.

4. Joubert's note: Book 3 of *Simple Remedies*, ch. 18.
This work is usually translated as *On the Properties of Simple Drugs*. Galen also gives another poison-immunity account, the famous story of Mithridates, in *On Antidotes* (I, i).

5. Joubert's note: Ian Lang. epist. 69. Book 1.
Johannes Lange (1485–1565) was a Silesian physician who wrote the *Medicinalium epistolarum miscellanea varia ac rara* . . . (Basel, 1554). After studying in Italy, he returned to serve as physician to four palatine electors. He advocated a return to the ancient texts.

6. Joubert's note: *Of Simple Remedies*, ch. 19.
See note 4 of this chapter above.

7. The poisonous and mythical effects of poppy, hemlock, and mandrake are well known. Hyoscyamine is an alkaloid derived from the seeds of the *Hyoscyamus niger* (*Oxford English Dictionary*).

8. Joubert's note: Book of Theriac to Pison. [*On Antidotes*, I, i and ii].

9. Claudius Aelianus was a third-century A.D. Roman who wrote two treatises in Greek, one on history, another on animals in seventeen books (to which Joubert is referring), *Peri zoon idiotetus*, commonly called in Latin *De animalium natura*.

10. *Vulue* did not have in the sixteenth-century the specific meaning of "vulva" that it has today. Based on the Latin *vulva* (womb, matrix), Joubert's term meant the vagina, as Cotgrave seems to confirm: "the wombe-pipe, or priuie passage; the way, or entrance into the wombe; also, the matrix, mother, or wombe itselfe."

11. Locusta (Lucusta) was known for concocting deadly compounds. She was supposed to have helped Agrippina in poisoning Claudius, and Nero in the death of Britannicus. She was put to death by the Emperor Galba for the evils she allegedly committed during Nero's reign.

12. Joubert's note: Dec. 2, Parad. 1.
See note 12 of chapter II in Book One above.

NOTES TO THE THIRD BOOK OF POPULAR ERRORS

BOOK THREE, CHAPTER I

1. Montluc was a commander of the Catholic armies. See note 6 of chapter I in Book Two above.

2. Civil wars: better known today as the Religious Wars.

3. I have translated the name *Porcelets* as "Piglets," as dictated by the context.

4. Giovanni Pico della Mirandola (1463–94), the celebrated Italian humanist and theologian, wrote numerous philosophical works in the Neoplatonic vein. He was an intellectual prodigy but died at only thirty-one years of age. Whether he was poisoned is still contested by scholars. His voluminous *Opera omnia* were published in Basel in 1573.

5. Albucasis (Abul Kasim) was a tenth-century A.D. Arab-Spanish physician. He wrote an encyclopedic medical work in thirty parts, of which the surgical tract, translated into Latin by Gerard of Cremona, was the primary manual on surgery until Guy de Chauliac's *Chirurgia magna* appeared in printed form in the fifteenth century. Latin translations of his work (known as the "al-Tasrif") appeared in Venice in 1497.

6. Pliny the Elder, of the first century A.D., wrote books on history and rhetoric, which have been lost. His *Historia naturalis*, in thirty-seven books, was frequently cited by writers of the Renaissance and is the source of many of Joubert's medical commonplaces. Joubert seems to be mistaken about the number of aborted children: Pliny mentions only seven (Book VII, iii, 33).

7. Marcin Kromer (1512–89) was a Polish humanist, theologian, and historian. The work and edition to which Joubert is referring is the *De origine et rebus gestis Polonorum libri triginta* (Basel, 1558).

8. We must not see in this reference a vulgar pun based upon the sixteenth-century French expression *mere des histoires* (Huguet), but rather a metaphoric use of the word *mer*. The *Sea of Histories* is an allusion to the fact that all history is derived from the "sea" of *tabularia* and *re gesta* of the reigns of various heads of state. Joubert cites chapter 16 of the second volume of the chronicles of Henry VII of Luxembourg.

9. Phthiriasis is "a morbid condition of the body in which lice multiply excessively, causing extreme irritation" (*Oxford English Dictionary*).

10. Joubert's note: Book 1, page 2.
See note 12 of chapter II in Book One above.

11. Joubert's note [in 1578 and 1579 editions only]: Gen. 17 & 21.

12. Joubert's note: Luke I[:36].

13. This is an instance of Joubert's respect for the Renaissance boundary between medicine and theology. See note 1 of chapter I in Book One above. Joubert nevertheless pursues the question on purely philosophical grounds (*enquerir par raison*).

14. Joubert wrote "four leagues" (*quatre lieuës*). See note 5 of chapter X in Book Two above.

15. Joubert's note: She was married twice, the second time to Tuech, and from her first husband had not had any children.

16. Although starting out as a mere barber-surgeon, Ambroise Paré (1509–90) remains one of the most important medical figures of the Renaissance. His celebrity began with his wound treatise, *La methode de traicter les playes faictes par hacquebutes et aultres bastons à feu* (Paris, 1545). In this revolutionary work, he contests dogmatic methods of treatment and lays the foundation for experimental medicine. Personal surgeon to Henry II, François II, Charles IX, and Henry III, he was widely known and respected for his surgical and obstetric techniques. His popularity grew even more with the publication of *Des monstres et prodiges* (1573) and his *Oeuvres* (1575). The fourth edition of this latter work contained the triumphant *Apologie*, which delivered the final blow to the dogmatism of the Renaissance physicians. See Dulieu for details of this debate.

17. Joubert's note: Book 7 of the *Hist. of Anim.*, ch. 4.

18. Joubert's note: Book 10, chap. 2.

The work of Aulus Gellius to which Joubert is referring is *The Attic Nights*. On Guy de Chauliac, see note 1 of chapter IV in Book One above.

19. Joubert's term is the *tunique Agnelete*, which Cotgrave translates as "Th'inmost of the three membranes which enwrap a wombe-lodged infant; called by some mid-wiues, the Coyfe, or Biggin of the child; by others, the childs shirt."

20. This sentence and the one preceding it appear only in the 1579 edition. Because of the internal time reference, we know it was written two years after the composition of the rest of the chapter.

21. Luigi Bonaccioli was an Italian physician of the late fifteenth and early sixteenth century. His work, the basis of his reputation, is known as the *Enneas muliebris* and is based upon two of his earlier publications: the *De uteri partiumque ejus confectione* (Strasbourg, 1537); and the *De conceptionis indiciis nec non maris foemineique partus significatione* (Strasbourg, 1538). It was later incorporated into a composite work entitled *Opus physiologicum et anatomicum de integritatis et corruptionis virginum notis*, (Lyons, 1639), which featured Séverin Pineau's *De integritatis et corruptionis virginum notis*, Bonaccioli's *Enneas muliebris*, Felix Plater's *De origine partium, earumque in utero conformatione*, Pierre Gassendi's *De septo cordis pervio observatio*, and Melchior Sebiz's *De notis virginitatis*.

22. I have used "teeny little women" to translate Joubert's *femmelettes* (Cotgrave gives "prettie little woman"). Joubert uses the term to designate small women, but the term borders on the familiar and the precious.

23. Joubert, contrary to his previous stance on the matter (see note 2 of chapter I in Book Two above), here makes more of a concession to the woman's part in the generation of the child. Indeed, Aristotle's theory does not offer a satisfactory explanation of a child's resemblance to its mother. Joubert is careful not to contradict Aristotle; rather, he maintains that several have misunderstood what the master says on the matter.

24. The phrase beginning " . . . large and well furnished . . . " and extending to the end of the sentence is lacking in the 1579 edition; it is present in all other editions. Its absence is thus either an oversight or a response to the

criticism following the 1578 edition. See note 1 of the Dedicatory Letter to Marguerite of France above and also the Introduction.

25. "Here lies Messer Concia, for having shot his bolt sixteen times plus one [a pound and an ounce]." *Chiavare* is an archaic verb meaning "to turn the key" or "to lock, to nail," and also, in a vulgar sense, "to have sexual relations."

26. This paragraph appears only in the 1579 edition.

27. Joubert's note: Concerning superfetation.

28. Joubert's note: Book 7 of the *History of Animals*, chap. 4.

29. Joubert's note: Galen, *De aig. matr.*, ch. 31.

BOOK THREE, CHAPTER II

1. Joubert's note: In the 6th Book of the *Hist. of Anim.*

The passage to which Joubert is referring is in section 18.

2. It is important to note that, when Joubert differs with Aristotle, he is careful to qualify his statements so as never to contradict him openly. Toward the end of this chapter, however, he expresses a clear difference with the firmly established notion of Hippocrates that children born in the eighth month are rarely viable.

3. Joubert's note [1579 edition only]: Vulgar people from Béarn say: "A whore and a sheep are always in season."

4. This is a frequent theme of Joubert's. See note 5 of chapter VII in Book One above.

5. The sections that follow constitute a robust example of Renaissance copia, the seemingly endless listing of items in a series. This device comes immediately after Joubert specifically states that his explanation is addressed to laymen.

6. Joubert's note: Book 4 of the *Gen. of Anim.*, ch. 10.

7. Joubert's reference is vague. Certain elements of his reminiscence fit the first tale of the seventh day of Boccaccio's *Decameron*, but not all. I have not been able to ascertain the title of this work, which Joubert calls the *liure des ioyeuses avantures*. Barbara C. Bowen suggests *Les ioyeuses aventures et facetieuses narrations* (Lyons, 1556).

8. *De natura deorum*, ii, 27.

9. Most likely because of an oversight of the compositor, the phrase "*. . . l'autre demain, &. . .*" ("another tomorrow, and") is lacking in the 1579 edition.

10. Joubert's note: L. Septime mense. ff de statu hom.

This is a reference to the *Digesta*, or the *Pandectae*, a collection of writings of Roman jurists, in fifty books, subdivided according to subject matter. The author Joubert cites is Julius Paulus, a third-century Roman lawyer who is responsible for about one sixth of the *Digesta*. Joubert uses the legal text abbreviations current during the Renaissance: l. = law; ff = *Digesta*.

11. Joubert's note: Book 7, ch. 5.

This work by Hippocrates (*Peri oktamenou*) is commonly known as *Eighth-Month's Child*. See note 7 of chapter I in Book One above.

12. *Historia naturalis*, VII, v.

13. Joubert's note: Ptolemy, *Centiloq[uium]. Propos[itio].* 51.

Ptolemy's influence on Renaissance medicine, philosophy, and astronomy was considerable. His complete works appeared in Basel in 1541, but many tracts were published much earlier. A Latin edition of the *Tetrabiblos*, in which figured the *Centiloquium* cited by Joubert, was published in Venice in 1484.

BOOK THREE, CHAPTER III

1. Joubert's *De urinis* is a twenty-five-chapter work on the color, odor, quality, and quantity of urine as symptom, using his own observations and many from antiquity, especially Galen. The work was included in his two-volume *Opera latina*, published in Lyons in 1582. See Amoreux, 63–90, and Dulieu, 151–62. Diagnosing pregnancy from urine, as Barbara C. Bowen notes, is a frequent comic device in the farce.

2. Cotgrave defines the word as follows: "Serositie; the waterishnesse, or thinner parts of the masse of bloud (answering to whay in milke) which floats upon it after it hath beene let out of a veine; also, the wheyish, or waterish moisture drawne by the kidneyes from all parts of the bodie, and after some concoction tearmed, Vrine."

3. The "pumping vein" is Cotgrave's translation of *veine emulgente*; Joubert used *vaisseaus emulgeans*. Cotgrave is more precise: "One of the two maine branches of the hollow veine [vena cava]; which goes to the reines [kidneys], and there is diuided into diuers others . . . "

4. Joubert's term for the spermatic vesicles is *parastates*; but it must not be confused with the prostate gland. Cotgrave defines *parastates* as "the conduits, or passages whereby the seed goes from the kidneys [loins] in the act of generation; or, two kernels which grow at the end of the bladder, and receiue the seed brought unto them by *Vasa deferentia* [*sic*] . . . "

5. Joubert's note: Objection. Response.

6. Joubert's exclamation quotes a common expression, *Balhe luy belle*. Cotgrave explains it at length "(of one that hath done, or spoken foolishly) faire befall him; let him euen haue it a Gods name. *Ironically*."

7. Joubert's note: Objection. Response.

8. Joubert's note [1579 edition only]: Equivocal signs of pregnancy.

9. Because of the printer's oversight, the 1579 edition lacks the phrase in parenthesis: " . . . lack of appetite (or appetite) for unusual or absurd things . . . "

10. Joubert's note [1579 edition only]: Univocal and absolute signs of pregnancy.

11. Joubert's note [1579 edition only]: A cow that has retained the sperm straightens up to such an extent that it arches its back upward. From this one is certain that it has conceived.

12. Joubert's note: *Aphorism* 41, Book 5.

Hippocrates's *Aphorisms* are in seven books. They were first translated into French by Martin de Saint-Gille in Avignon between 1360 and 1365.

See Germaine Lafeuille, *Les commentaires de Saint-Gille sur les 'Aphorismes Ypocras'* (Geneva, 1964). François Rabelais published an edition of them in 1532.

BOOK THREE, CHAPTER IV

1. Joubert's note [lacking in the 1579 edition]: *Aph.* 42, Book 5. See note 12 of the preceding chapter.

2. Joubert's note [lacking in the 1579 edition]: *Aph.* 48, Book 5.

3. By this Joubert means that any other complicating conditions must be put aside when one is theorizing.

4. Cotgrave gives the following definition of this word: "A Timpanie, or Moone-calfe; a shapelesse lump of flesh, or hard swelling, in the wombe, that makes a woman seeme withchild . . . "

5. Cotgrave's definition of *Harpye* is also worthy of note: "An Harpie; one of those monstrous, and rauenous birds which Poets faine to haue had womens faces, hands armed with talons, and bellies full of an ordure wherwith they infected all the meat they touched; and hereof are the monstrous brood (Moles, or Moone-calues) of some women so tearmed."

6. Joubert's note: Chapter 7.

7. Joubert thus ends on a humorous note a discussion that embraces empiricism. It is important, however, in order to maintain a balanced view of Joubert's thought, to recall the numerous instances in which he warns against the danger of uninformed practice. See, for example, note 3 of chapter III and note 4 of chapter XIX in Book One above, and note 23 of chapter I and note 2 of chapter II in Book Three. See also his espousal of Galen's plea for a theoretically grounded practice (note 1 of chapter XVII in Book One above).

BOOK THREE, CHAPTER V

1. Joubert's term is *Pie & Mollesse.* Cotgrave gives roughly the same definition for both: "*Pie* . . . also, the monstrous appetite of maides, and big-bellied women, unto Coales, Ashes, Paper, and such other unnaturall meats . . . "; "*Mollesse* . . . also, the monstrous appetite of some maids, and women, unto paper, ashes, coales, and such other harsh, and unsauorie acates [things purchased, provisions, foods]."

2. Aetius was a physician of Mesopotamia of the fifth and sixth century A.D. His work was entitled *Tetrabiblos* and contained a compilation of writings of earlier physicians and his own observations on pathology and surgery. Joubert could have consulted one of the Latin versions of the translation done by Janus Cornarius, the *Contractae ex veteribus medicinae tetrabiblos,* published in Venice in 1543.

Paul of Aegina was a Greek medical writer of the seventh century A.D. He wrote the *De re medica libri septem,* drawing upon Galen and other sources.

Rhazes (A.D. 850–923) was a famous Arab physician often cited in medical texts of the Middle Ages and Renaissance. His treatise on smallpox and measles was well known. Robert Estienne published the Greek version of his treatise in 1548, with the corrections of Jacques Goupil; Sébastien Colin published it in French translation in 1556.

For Avicenna, see note 6 of chapter II in Book One above.

3. Joubert blames laymen so as not to confront in open opposition the church, the true source of the layman's fear of sharing responsibility for the sin of abortion, which carried the sanction of excommunication. Numerous church councils had condemned it: Elvira, 313; Ancyra (today known as Ankara), 314; Lérida, 524; Constantinople, 706. Pope Sixtus V (1520–90) published in 1588 a papal bull (*Effraenatam*) reiterating the church's opposition to the practice.

4. Bloodletting was the standard treatment for severe fever (Hippocrates, *Epidemics*, III, case viii). But Hippocrates also forged the link between bloodletting and miscarriage in his *Aphorisms* (V, xxxi): "A woman with child, if bled, miscarries; the larger the embryo, the greater the risk."

5. Both Galen and Hippocrates prescribed violent purgatives, such as hellebore (Galen, *On the Natural Faculties*, III, xiii; Hippocrates, *Aphorisms*, IV, xiii–xvi, and in the first book of the Hippocratic treatise entitled *Regimen in Acute Diseases*). See also Celsus, *De medicina*, II, 12.

6. Cotgrave gives the proverb: "*Se remplir de foin, ou de paille.* To stop his guts with, to glut himselfe on, anything; (from a horse, that wanting Hay falls to his Litter)."

BOOK THREE, CHAPTER VI

1. Joubert's term is *orgeol*; Cotgrave translates it along the lines of Joubert's definition: "A long wart resembling a Barlie corn, and growing on the edge, or corner of an eye-lid."

2. The French dialect frequently characterizing Joubert's vocabulary is that of Languedoc.

3. The modern French term is *orgelet*.

4. Joubert is referring to his *Treatise on Laughter*, Book One, chapter XXVII, 61–62.

5. The Volsci were a people in the south of Latium. Joubert must be referring to the *Livre doré de Marc-Aurele*, because this anecdote does not appear in Marcus Aurelius (neither in his *Thoughts* nor in his correspondence). The story is told in Plutarch's *Lives* ("Camillus"), and aspects of it may have been drawn from Livy (v and vi). Joubert's confusion is understandable: both Plutarch (A.D. 50–125) and Marcus Aurelius (A.D. 121–80) wrote in Greek and were roughly contemporary.

6. All editions number in the margin these five privileges.

BOOK THREE, CHAPTER VII

1. This paragraph appears only in the 1579 edition.

2. Joubert's note [lacking in the 1579 edition]: Maybe he did not cut anything but granted her the favor of the flesh hanging between his thighs.

3. On morphew, see note 5 of chapter III in Book Two above.

4. Joubert's term is *loupes*; Cotgrave translates it as a "flegmaticke lumpe, wenne, bunch, or swelling of flesh under the throat, bellie, &c; also, a little one on the wrist, foot, or other joynt, gotten by a blow whereby a sinew being wrested rises, and grows hard . . . "

5. Anne de Joyeuse (1561–87), a favorite of Henry III, was an admiral of France and leader of the Catholic armies during the Religious Wars (1562–93).

6. Joubert misquotes Hippocrates slightly: Hippocrates allows seventy days for movement. Joubert is citing section 42 of *Nutriment* (*Peri trophes*).

7. Women's cravings, according to Joubert, are the result of the retention of the menses and are thus explained in terms of the theory of the four humors. See chapter V in Book Three above.

8. This sentence appears only in the 1579 edition.

9. See note 6 of chapter VI in Book One above.

BOOK THREE, CHAPTER VIII

1. Joubert takes up once again his metaphor of baking bread in an oven to describe the condition of the child in the womb. See chapter II of Book Three above.

2. Joubert is referring to Galen's *De temperamentis* (*Peri kraseon*), which would have been available to him in the form published by the Parisian printer Jean Dupuys in 1554.

3. Cotgrave translates the expression as follows: "*Plus estourdi que le premier coup de Matines.* More drowsie, and amazed, then a Frier at the first toll of the Morn-sacring bell."

4. This point was to be made famous by Montaigne in his *Essais* ("De l'institution des enfants" I, xxvi; and "Des menteurs," I, ix).

5. Renaissance Latin often rendered the *ae* ending phonetically as *e*. Thus, *uve passe* should read *uvae passae*.

6. This work by Joubert was never published and has unfortunately been lost. See Amoreux, 105.

BOOK THREE, CHAPTER IX

1. On the renewed interest in anatomy during the Renaissance, see Brabant, 211–71. Joubert will modify his original claim of the stupidity of this popular saying in the course of the chapter.

2. These foods were apparently considered nonessential in the Renaissance diet. Huguet, citing Montaigne and Cholières, suggests a possible explanation of the meaning of "Spanish foods" as either finely butchered meats or foods lacking in substantial nourishment.

3. It can be seen here that another of Joubert's purposes in writing the *Erreurs populaires* was to account for those practices of laymen that were indeed beneficial, but of which their understanding was incorrect—another error Joubert felt compelled to expose.

NOTES TO THE FOURTH BOOK OF POPULAR ERRORS

BOOK FOUR, CHAPTER I

1. This is among the most interesting chapters from the point of view of the corrective posture assumed by Joubert. It was not until Vesalius gave the first accurate description of the pelvis in 1543 that Renaissance anatomists and physicians began asserting that the pelvic bones did not separate at the birth of the child. Joubert is thus following Vesalius's demonstration of the fallacy of the notion of the gaping of the pelvis, and the fact that it formed, rather, an unyielding bony ring. The doctrine of the separation of the pelvic bones was so tenaciously held that it was even promulgated by the sagacious Ambroise Paré. We should not fail to note that Joubert takes great pleasure in contradicting Paré on this point; the barber-surgeon, already enjoying increasing success and celebrity because of his successful surgical techniques, was also becoming famous for his writings following the publication of his vulnerary tract, *La methode de traicter les playes faictes par hacquebutes et aultres bastons à feu* (Paris, 1545).

2. Women without shame.

3. Joubert's source for the wisdom of Lycurgus, the ninth-century B.C. Spartan lawgiver, is Plutarch's *Lives*.

4. *Kokkyx* in Greek means "cuckoo."

5. The *Thesaurus linguae latinae* gives this form as *curuca*: " . . . *linosa vulgo avis quae dicitur [curuc]a*." Cotgrave translates the word Joubert used (*Verdalle*) as "dunneck [dunnock], or Hedge-sparrow." It is possibly the *Prunella modularis*.

6. *Becco*, in Italian, can mean "beak" or "he-goat"; it can also mean "cuckold."

BOOK FOUR, CHAPTER II

1. See note 1 of the preceding chapter.

2. Joubert's note: *History of Animals*, Book 7, chapter 4.
This reference does not apply to the matter under discussion. It is misplaced and in all likelihood meant to accompany the passage referred to in note 5 below.

3. The French word for midwife is *sage-femme*, literally, a "wise woman."

4. This last section of the chapter could be read as a surreptitious attack on the ability of midwives to cope with multiple parturitions, especially in view of Joubert's scorning of matrons in the next chapter.

5. These stories of exposure all follow the same pattern as that set down in Marie de France's *lai* entitled "Le fresne": fear that multiple births inevitably entail accusations of adultery, a notion supported by Aristotle in the *Historia animalium* (VII, 4).

6. The Seigneur d'Estourneau seems to be a very obscure figure for one

who was the maître d'hôtel of Henry IV, as Joubert claims. A certain Estourneau is mentioned in a letter of Catherine de' Medici (16 December 1571), where he is called a *commissaire*; otherwise, I have not found his name in the documents of the time that were at my disposal. Could he possibly be the son of Jacques-Matthieu d'Estourneau, the architect of Châteauneuf-sur-Sarthe?

BOOK FOUR, CHAPTER III

1. The physicians generally relinquished surgery to the surgeons and barber-surgeons during the Renaissance.

2. See note 2 of chapter VII and note 3 of chapter XI in Book One above. Joubert could well be alluding to Louyse Bourgeois; see note 5 of chapter IV in Book Five.

3. Guillaume Rondelet (1500–1584), called the father of ichthyology, was an anatomist, rector of the school of medicine in Montpellier, and Joubert's major professor. He also wrote the *De piscibus marinis libri XVIII* and the *Universalis aquatilium*, which were translated into French a few years after their publication in 1554. He was immortalized in Rabelais's *Tiers Livre* (xxxi–xxxiii) under the name of Rondibilis.

BOOK FOUR, CHAPTER IV

1. The reference is to the ligaturing and cutting of the umbilical cord and the length to be left.

2. Hippocrates had established the link between cold and several illnesses, among which figured diarrhea (*Airs Waters Places* [*Peri lepon udaton topon*], III).

3. Joubert was never to write this book (see the Introduction), and so we have no further details of this disease. It may have been an infection of maggots; Celsus only mentions them occurring in the ear (*De medicina*, VI, 7.5). Celsus also mentions worms, both round and flat, exiting from the bowels and the mouth, as does Hippocrates (*Prognostic*, XI), but not from the spine (IV, 23.24).

BOOK FOUR, CHAPTER V

1. On Avicenna, see note 6 of chapter II in Book One above; on Rhazes, see note 2 of chapter V in Book Three above.

2. Little is known of Antonio Guainiero (Guainerius) other than his publications, many of which appeared without an indication of date or place: *Antidotarium* (1472), *De iuncturis*, *De passionibus stomachi*, *De pleuresi commentaria* (1483), *Tractatus de matricibus* (1484), *Practica* (1488). A collection of his works was published in Lyons in 1525. The fact that Guainiero is a minor figure of the previous century allows Joubert to contradict him more vigorously than the Arab physicians maintaining the same opinion.

3. Joubert is referring to chapters XVI and XIX of Book One.

4. Joubert's diatribe against women puts distance between the the Arab physicians' notion of the umbilicus as trustworthy sign of future offspring and his own refutation of this notion.

5. Joubert's paraleipsis, although effective, is not meant to be as destructive as it might seem. He has always maintained that arguments should not be tainted by abnormal circumstances or accidents. This very point is emphasized in chapter I of Book Three. In the last passage of the present chapter, Joubert reasons very much à la Montaigne as he undoes the link between foreknowledge and natural phenomena, a link whose strength remained generally vigorous throughout the Renaissance.

6. This is possibly a reference to Celsus's advice on incisions in *De medicina* (VII, 19).

BOOK FOUR, CHAPTER VI

1. That is, still surrounded by the amnion.

2. Joubert erroneously gives *amnie* as the Greek form. He may have been attempting a feminine form (*amnia*, or *amne*) so as to coincide with the feminine French form *Agnelette*, which calls to mind the supposed source of the word, lambskin (*amneios*, or *amnios*).

3. Joubert is playing on the words *monde* ("world") and *immonde* ("unclean"), a common pun at the time, and noted by Cotgrave in the following proverb: "*Qui veut la conscience monde, il doit fuïr le monde immonde*: *Prov*. He that affects a cleane conscience, must auoid unclean copemates."

4. The literal translation ("Amongst shit and piss the handsome son takes his nourishment") fails to transmit the rhythm and rhyme of the Languedoc dialect.

5. Grimache, known as Maître Grimache, is an obscure literary figure of the French Renaissance who wrote *Le vrai medecin qui guerit de tous les maux*. See Sabatier, 257.

BOOK FOUR, CHAPTER VII

1. *Aeneid*, iii, 210–18; 250–52; 330–36. For Cotgrave's definition of the Harpy, see notes 4 and 5 of chapter IV of Book Three above.

2. For a fuller version of the story of Phineus and the Harpies, see Apollonius Rhodius (*Argonautica*, ii, 178–300).

3. Joubert's note [1579 edition only]: Canto 33 [96–128].

Ludovico Ariosto (1474–1533) was an Italian humanist and poet, whose great epic poem, *Orlando furioso*, had an immense influence on French literature of the Renaissance.

4. Niccolò de' Niccoli, an Italian humanist of the fifteenth century, was known for the brilliant luxury in which he lived, and for his accumulation and critical editing of manuscripts in Florence. Joubert must be referring to some unpublished piece by Niccoli, for he left no written work. Bernard de Gordon was a French physician of the late thirteenth and early fourteenth centuries. Known as the *fleur de lye* [*sic*] *en medecine*, he taught at Mont-

pellier and wrote many works which were gathered and published in Venice in 1498, in Paris in 1542, and in Lyons in 1559: *De regimine acutarum aegritudinum; De signis prognosticis; De urinis et cautelis earum; De pulsibus.* His *De conservatione vitae humanae*, his *De phlebotomia*, and his *De floribus dietarum* appeared in Lyons in 1580. Other titles attributed to Gordon are the *De victus ratione et pharmacorum usu in morbis acutis*, the *De crisi et criticis diebus, atque prognosticandi ratione*, the *De medicamentorum gradibus*, the *De marasma*, and the *De theriaca*.

5. Besides Agrippa d'Aubigné's active role in the Religious Wars under Henry of Navarre, he is known today for his writings: *Les Tragiques*, a vigorous epic recounting the French civil-religious struggle, the *Histoire universelle*, a prose recounting of numerous Religious War episodes, and the *Printemps*, a sequence of love sonnets. For a discussion of D'Aubigné the writer and this anecdote of the *Erreurs populaires*, see G. de Rocher, "Agrippa d'Aubigné's Erinyes: The History of a Field of Poetic Imagery," *Renaissance Papers 1981*, 45–54.

6. On Philibert Sarazin, see the standard biography of D'Aubigné: A. Cavens, *Agrippa d'Aubigné: l'homme et l'oeuvre* (Brussels, 1949).

7. The alleged case of the Châtellerault woman appears only in the 1579 edition. It came perhaps to be an embarrassment to Joubert and was later removed. He was obviously convinced at one time of its anecdotal strength, if not its veracity, and internal evidence suggests knowledge of a follow-up inquiry, as well as a certain sympathy with the case.

8. The word "flesh" (*chair*) is lacking in the 1579 edition. Aristotle discusses the *mola uteri* in *De generatione animalium*, IV, vii.

BOOK FOUR, CHAPTER VIII

1. On Joubert's family, see Amoreux, 19 and 107–11, and Dulieu, 139–40 and 159–60. According to Cotgrave, "In France the title of *Chevalier* is, often, a bare title of honour, and ordinarily conferred on great Officers (whether of the short, or long, Robe,) and on the Lords of great, and meane Seignories."

2. Joubert numbers each sibling in the margin. The meticulously systematic listing of evidence is noteworthy, especially by a physician who usually attacks empirical applications. See the Introduction and note 7 of chapter IV in Book Three above.

3. That is, during the last quarter.

4. Joubert does not give Magdaleine's specific time of birth.

5. The ephemerides were astronomical almanacs published each year. If we trust the two fragments in the Bibliothèque Nationale, Rabelais published such almanacs; it is certain that he published parodies of them, as can be seen by his *Pantagrueline prognostication certaine, veritable, et infaillible* . . . from 1532 to 1542. See Rabelais, II, 500–524.

6. That is, the proposition of the chapter title.

7. On Joubert's wife, see Dulieu, 159–60.

8. Many conclusions can be drawn from this last phrase adding Joubert's

sixth child, Henry, to the list of his children. The most important, without a doubt, is that it fixes a *terminus a quo* for at least one of Joubert's revisions of the *Erreurs populaires*: June 6, 1577 (two years before they appear in print). This suggests that the first edition, published in 1578, was already obsolete and incomplete: excisions, additions, or corrections could no longer be made. Response to the sharp criticism of the first edition, therefore, could only be hasty and spotty, for the expanded second edition would have already been in composition (both in Paris and in Bordeaux, as it happened), further complicating any systematic revision. It is obvious from an examination of the various editions that printers were driven by one overriding concern: to produce another edition, or even another printing, of the *Erreurs populaires* as soon as possible to meet demand. Finally, it made of the 1579 edition a sort of *hapax legomenon*, in that later editions, for a variety of reasons (see the Introduction), came to be based upon the 1578 edition.

9. In accordance with what is discussed in the above note, the phrase "and Henry" (*& Hanry*) appears only in the 1579 edition.

10. These last two paragraphs of the chapter are also additions and appear only in the 1579 edition.

11. Joubert is referring to the second chapter of Book Three above. His repetition of his belief in the powerful influence of astrology emphasizes the important role the science played in Renaissance medical theory.

BOOK FOUR, CHAPTER IX

1. Hippocras, named after Hippocrates, was a spiced and honeyed wine.

2. I am grateful to Barbara C. Bowen for pointing out the relevance of this passage to today's breathing exercises for pregnant women.

3. Celsus warns that cold is harmful to the womb (*De medicina*, I, 9.3).

4. I have thus translated Joubert's "*gros boyaus, nommés Colon & Cullier*," because *cullier* was a common term in use at the time to designate the rectum. See Screech's remarks on vulgar terms in the language of sixteenth-century France (*Rabelais*, 51), and Amoreux, 36–45.

5. The rule of women in matters of childbirth and lying-in was duly respected throughout the Renaissance. See Gordon, 686–95, and Davis, 145, 261, and 313, n. 37.

6. Acute diseases were defined by the presence of high fever, but if their duration exceeded twenty days, they became chronic (Hippocrates, *Regimen in Acute Diseases* [*Peri diaites oseon*], v; Celsus, *De medicina*, III, 1–2).

BOOK FOUR, CHAPTER X

1. The notion of *horror vacui* (*kenoumenon*), first developed by Erasistratus (a physician of the third century B.C.), was taken up by Galen, one of Joubert's sources (*On the Natural Faculties* [*Peri physikon dynameon*], I, xvi; II, i and vi).

2. Joubert uses the phrase *bien celebrée*; this confirms the importance of the lying-in period, as emphasized by Davis and Thomas (see note 5 of

the preceding chapter). Cotgrave defines *celebré* as "celebrated, or solemnized with great assemblies; also, magnified, renowmed [*sic*], exceedingly honoured." See also Joubert's remarks about the joy of the parturient mother in chapter VI of Book Two.

BOOK FOUR, CHAPTER XI

1. "Which" (*qui*) is lacking in the 1579 edition.
2. See note 3 of chapter III in Book Three above.
3. Joubert's note: Objection [followed by] Solution.
4. An aposteme was an abcess; Cotgrave defines "Aposteme. *Seeke* Apostume. . . . *Apostume*: *f.* An Impostume; an inward swelling full of corrupt matter."

BOOK FOUR, CHAPTER XII

1. Joubert has collapsed what had earlier (chapter IX) been two separate points. Cold penetrated the womb of the parturient mother because of improper care, not heavy blood.
2. Joubert's peroration seeks to refute this popular notion based on the divine sanction in Genesis 1:16. He adopts an opposing attitude: the insatiability of women, already developed in the first chapter of Book Three.

NOTES TO THE FIFTH BOOK OF POPULAR ERRORS

BOOK FIVE, CHAPTER I

1. Joubert's note: Book 12, ch. 1.

The citation is correct. Favorinus, cited by Joubert as an Athenian philosopher, was actually a second-century A.D. philosopher born in Arles. Aulus Gellius, a Roman writer of the second century A.D., left twenty books of *The Attic Nights* (*Noctes Atticae*), which are essays based on his readings of Greek and Latin texts. He was widely read during the Renaissance.

2. Joubert's note: Imprecation or cursing in the Languedoc dialect, where in Italian one would say *Checancaro* [*Que canchero*], and in French *Que diable*.

Rabelais uses the expression in the Prologue to *Gargantua*: " . . . *que le maulubec vous trousque* . . . " (I, 9). Cotgrave translates *mau-loubet* as "The Wolse (a disease)"; *The Oxford English Dictionary* gives "an ulcerous disease of the skin, sometimes erosive, sometimes hypertrophous." The equivalent English expression at the time was "A pox on . . . "

3. *Aeneid*, iv, 365–67. Favorinus (Aulus Gellius?) cites Dido's angry outburst at Aeneas when he tells her he must leave for Italy. The reference to Homer is in the *Iliad*, xvi, 33. This passage shows Joubert leaning once again toward the Galenist position of female sperm.
4. Antonio de Guevara (1480?–1548?) was a Spanish historian who be-

longed to the Franciscan order. His famous work, the *Libro aureo de Marco Aurelio* (Valladolid, 1529), was translated into French (both as the *Livre doré de Marc Aurele* [Paris, 1531] and as *L'Orloge des princes* [Paris, 1540]) and English (*The Dial of Princes* [London, 1577]). Although challenged for its veracity, it was widely read, went through numerous printings and editions, and was much used by Montaigne. As Joubert indicates, it is the source of nearly all the citations and anecdotes of this chapter.

5. Joubert's note: Book 3 of *Laws*.

The allusion may be to Cyrus's upbringing (694 D) as contrasted with that of Darius (695 B–D). See Guevara, *Dial of Princes*, II, xix.

6. Joubert's note: After the 10th *Parad.* of the first decade.

See note 12 of chapter II in Book One above.

7. Plutarch mentions the rumor that Alcibiades was nursed by a Spartan woman. Plato is mentioned in the same passage in connection with Alcibiades' preceptor, Zopyrus (*Lives*, "Lycurgus").

8. Joubert returns to this idea later in the chapter.

9. The preceding two paragraphs are drawn nearly verbatim from Guevara, *Dial of Princes*, II, xix.

10. Joubert's version of the story of Cornelius Scipio Asianus is from Guevara, who gives as source a supposed work of Sextus of Chaeronea entitled *De ambigua justitia*. It is curious that Joubert uses the expression *soeur de lait* (foster sister) and then has Scipio refer to her as a wet nurse in the next sentence. The error may spring from Guevara's attribution of the intercession to a *soeur dudict empereur*, which Joubert could have transcribed as *soeur de lai[c]t*.

11. Again, the source of the remarks concerning Antipater and Nero, as well as those regarding the Gracchi brothers, are from Guevara (*Dial of Princes*, II, xix). Guevara cites the "historiographers" for his knowledge of Antipater and Nero, probably referring to Livy and perhaps to Appian. The Gracchi episode comes from a supposed work by Junius Rusticus, *Bringing Up Children* (after Plutarch's *De liberis educandis*).

12. Joubert mistakenly wrote Thomistes for Othonistes. Also, according to Guevara, Othonistes was the sixth king of the Lacedaemonians. Guevara gives as source of this story Plutarch's *Reign of Princes*. Presumably, he meant the *De liberis educandis*, which is cited elsewhere (see note 15 below); but these names do not appear therein. Could he have meant the *Regum et imperatorum apophthegmata*, or possibly one of the *Moralia* dealing with statecraft: the *Maxime cum principibus philosopho diserendum*; *Ad principem ineruditum*; *An seni respublica gerenda sit*; *Praecepta gerendae reipublicae*; or the *De unius in republica dominatione, populari statu, et paucorum imperio*?

13. The suspicion alluded to is that of another child being substituted for the heir. This is the first danger cited by Jacques Guillemau when he advises against having the child nursed by a woman other than its mother. A student of Ambroise Paré, Guillemau was a French barber-surgeon who wrote one of the most popular works on obstetrics of the time, *L'heureux accouchement des femmes* (Paris, 1609). It was immediately translated into English as *Child-birth or, the Happy Deliverie of Women* (London, 1612). It is a

practical manual, filled with recipes for various applications, and not given to theoretical discussions, as evidenced by Guillemau's phrase: "But I will leaue this curious speculation to Physitions, and will onely meddle with that, which belongs to Chirurgions . . . "

Guevara, Paré, and Joubert were not alone in opposing the antinursing tradition; Barbara C. Bowen reminded me that Erasmus, early in the sixteenth century, had already advocated nursing one's own child.

14. Guevara, *Dial of Princes*, II, xix.

15. Guevara again cites Plutarch as his source (*De liberis educandis* [*Peri paidon agoges*], 5).

BOOK FIVE, CHAPTER II

1. Joubert gives *colostre* in all editions. A possible explanation for this form is that it is a phonetic spelling for *colostrae* plural of *colostra* used by Pliny.

2. Joubert's note: Book 11, chap. 41 & Book 28, chap. 9.

The 1587 edition omits the second reference. Both of these references to the *Historia naturalis* are inaccurate. The first should be XI, xcvi: here Pliny speaks of the lethal nature of milk for foals who take it during the first two days, and he gives the name of the disease (*colostratio*) caused by this milk. The second should read XVIII, xxxiii, in which Pliny says that colostrum is the first milk, "thick and spongy," after delivery.

3. Cotgrave defines this term as "She that giues sucke to a new-borne child for two or three days, and untill the mothers milke be fit for it: *Provençal.*"

4. This paragraph only appears in the 1579 edition. The Anabaptists sprang from the Swiss Reformation in Zurich, which began in 1523. Besides proclaiming a sharp division between church and state, they maintained that only confessed Christians could be baptized. The sect also practiced nonintervention in medical matters.

5. All the other editions give "It is true that . . . " instead of "Likewise . . . " ("*Il est vray que* . . . " instead of "*Samblablement* . . . "). The addition of the preceding paragraph necessitated the replacement of the former phrase with the latter to provide a proper transition.

6. This is an allusion to the belief that expectant women who nursed not only risked giving the infant bad milk made from retained menstrual blood but also deprived the embryo of necessary blood because the mammary glands were drawing too heavily on the available supply. See, for example, Pliny, *Historia naturalis*, XXVIII, xxxiii; Guillemau, *The Nursing of Children* (London, 1612), p. 3; and Hippocrates, *Aphorisms*, V, lii: "When milk flows copiously from the breasts of a woman with child, it shows that the unborn child is sickly; but if the breasts be hard, it shows that the child is more healthy."

7. Isaiah: 7:14–15.

8. On aurotherapy, see Brabant, 153–68. Brabant also discusses theriaca

(205–6) and mithridate (106–7 and 205–7). Galen gives an account of these preparations in the opening chapters of *De antidotis*.

9. The argument alluded to here is that since organ functions all but cease during sleep, any delays to be observed between ingestion or application of remedies must exclude time during which the patient slept. See Hippocrates, *Regimen* (*Peri diaites*), II, lx.

BOOK FIVE, CHAPTER III

1. Joubert's note: *Apho*. 39, Book 5.

See note 12 of chapter III in Book Three and note 7 of chapter I in Book One above.

2. That is, in comparison to the male sex, which, according to the theory of humors, is hot and dry.

3. Joubert's note: See the 1st chap. of the 3rd book.

4. Joubert's note: Book 7 *Hist. of Anim*, chap. 2.

Joubert refers especially to the opening sentence (582 B): "The onset of the catamenia in women takes place towards the end of the month; and on this account the wiseacres assert that the moon is feminine, because the discharge in women and the waning of the moon happen at one and the same time, and after the wane and the discharge both one and the other grow whole again" (D'Arcy Wentworth Thomas translation, Oxford, 1910).

5. Joubert, doubtless hesitant to mention him by name, is alluding to Aristotle, *De generatione animalium*, IV, viii (776 B 30).

6. Joubert refers here to Empedocles and his followers. Ibid., 777 A 9–11.

7. On the notion of vapors, see note 2 of chapter VIII in Book Two above.

8. Joubert's note: Book 4, *Hist. of Anim*, chap. 20.

The reference, incorrect, should read III, 20.

9. Joubert is citing the *Digesta*. See note 10 of chapter II in Book Three above.

10. This final paragraph appears only in the 1579 edition. One could argue for a shift in Joubert's theory toward a more empirically grounded view were it not for the last sentence in which he discounts the evidence by calling the occurrence a miracle. This allows him to dismiss it because it is a special case. Joubert accounts for milk in the breasts of virgins by having recourse to deductive medical reasoning (in this case, the concept of abundance), which would take precedence over authority or experience. He thus shuns empiricism, even in his last writings.

BOOK FIVE, CHAPTER IV

1. Joubert added this paragraph to the 1579 edition, probably in response to the sharp criticism he received regarding such passages in the 1578 edition. It is not found in the other editions.

2. That is, it is better for these women to gain the knowledge afforded through anatomy by reading than by actually witnessing a public dissection.

3. We see here the principal source of criticism launched against this section of the *Erreurs populaires*: the midwives, taken to task already in chapter XVI of Book One, chapter X of Book Three, chapters II, III, IX, and X of Book Four, and especially in the present chapter.

4. Joubert's note: *Aph.* 1, Bk. 1.

This aphorism is among the most often quoted: "Life is short, art long, opportunity fleeting, experience treacherous, judgment difficult. The physician must be ready, not only to to his duty himself, but also to secure the cooperation of the patient, of the attendants, and of the externals." The fable Joubert next recounts from Aesop is No. 166 (*Fabulae Aesopicae collectae*, ed. Teubner, Leipzig, 1863).

5. Joubert plays upon the term *sage-femme*. See note 3 of chapter II in Book Four above. In spite of Joubert's praise, Massale's popularity in Renaissance France was not as widespread as that of Louyse Bourgeois, for example. See Hurd-Mead, 356–57.

6. In the 1579 edition Joubert changed the word *leuandieres*, the official term for midwives, to *matrones*, a more general term, thus taking less direct aim at the midwives. See note 3 of this chapter.

7. The signs of virginity are numbered in the margin. Arabic numeral 1 is missing, however, in the 1579 edition, as is the Roman numeral I in the 1587 edition.

8. Joubert invokes Aristotle's four causes: formal (remote), material (adjacent), efficient (immediate), and final (ultimate), as discussed in *Metaphysica*, I, iii.

9. Cotgrave gives the meaning of this verb: "To grow hairie about the priuities; also, to be as lasciuious, or smell as ranke as a goat . . . "

10. Martial (first century A.D.) enjoyed considerable popularity in sixteenth-century France, beginning with Clément Marot's *Epigrammes* and culminating in Montaigne, who appreciated Martial's wit, sting, and obscenity. Joubert is referring to epigram xxii of Book XI (ll. 7–8).

11. *Rumex patientia*.

12. Joubert's note: Novella 10, day 3.

This reference to the *Decameron* was not criticized by Joubert's contemporaries as obscene.

13. Marguerite de Navarre, *Heptameron*, Third Day, second tale.

14. This phrase recalls Aristotle, *Metaphysica*, II, i.

15. Aristotle, *Metaphysica*, IV, vi (1011 B 15).

16. Each of the following signs is numbered in all the editions. Because Joubert criticizes these terms, with the exception of *dame du milieu* and *reffiron* (or *arrierefosse*), as jargon peculiar to midwives and known only to a member plying the trade, I am not attempting to translate them. Cotgrave, although containing virtually all of them, does not furnish an English equivalent; this is understandable since, on the one occasion he does, he gives "whimwham" and "gewgaw." Each of the three depositions, therefore, keeps the original terms for the specific parts of the genitalia.

17. Again, each of the signs is numbered.

18. *Os Bertrand*, also called *os sans nom*, is the *os pubis*. See note 1 of chapter I in Book Four above.

19. Jean Fernel (1496–1558) was a renowned physician who served as Henry II's physician and was credited with curing the sterility of Catherine de' Medici. He wrote an astronomical treatise (*Cosmotheoria* [Paris, 1528]) and several medical works, among which figures an important work on syphilis, published after his death, *De luis venereae curatione perfectissima liber* (Antwerp, 1579).

Jacques Dubois, or Sylvius (1478–1555), was an illustrious professor of medicine in Paris. The sixteenth-century storyteller Noël Du Fail has immortalized Dubois, recounting episodes of the professor's onomastics of the genitals, malodorous dissections, and his unrelenting avarice.

Jean Vassès de Meaux, or Vassaeus (1486–1550), wrote several medical works on Galen and Hippocrates. After serving as regent, he became dean of the Faculté de Médecine in Paris in 1522.

20. The passage in parentheses was suppressed in the 1579 edition. See note 1 of the Dedicatory Letter to Marguerite de France and also the Introduction.

21. One is tempted to read Joubert's term for a woman's particular genital structure (*conformation*) as a pun involving the vulgar word for the female genitalia (*con-formation*).

22. "In conjunction with" (*ioint*) appears to be another play on words based upon the confusion of the literal and figurative meanings of the conjunction.

23. "By the shape of the nose the *ad te levavi* is known." *Ad te levavi* was a vulgar scriptural pun used to designate the penis: "To you I have lifted up my . . . " (Psalm 123). It appears, for example, in Rabelais (*Gargantua*, XI).

24. I have used the archaic form "privities" to translate Joubert's *cas* in order to preserve the legal overtones of this sixteenth-century term for the genitals.

25. The sharp criticism of the scabrous nature of the first edition of the *Erreurs populaires* may well have led Joubert to change *manche* to *membre* in the 1579 edition; *manche* was a more colloquial term (Cotgrave gives "a mans toole") than the more proper *membre*.

26. See the preceding note.

27. The 1579 edition changed *passage* to *conduit* for the reason mentioned in note 25 above. The argument that follows recalls that of Saint Augustine (*City of God*, I, xviii), in which a clumsy midwife accidentally ruptures the hymen in attempting to verify a woman's virginity. Augustine's anecdote is cited by Montaigne in the third book of his *Essais* (III, v).

28. The 1579 edition tightens the syntax of this sentence from (1578) "*. . . non moins que a un' autre, qui . . .*" to "*. . . non moins que à celle qui s'ansuit, laquelle ne faut oublier.*" Unfortunately, inasmuch as the later printings and editions are all based upon the first edition, the emendation was not retained.

29. The Belgian anatomist Vesalius (1514–64), famous for his masterpiece, *De humani corporis fabrica libri septem* (1543), also published other anatomical and surgical treatises. He is credited with having produced the first critical medical illustrations and thereby contributing greatly to experimental medicine. Vesalius's disciple, Gabriello Fallopio, or Fallopius (1523–62),

was a celebrated Italian physician, surgeon, and anatomist known as "the Aesculapius of his century." He succeeded his master at Padua, where he wrote his *Observationes anatomicae* (Venice, 1561) and composed the drawings of his *De formatio foetu* (not published until 1604). His anatomies allowed him to free himself from the oppressive influence of Galen and to describe accurately the internal ear, the head, and the genitals. His collected works, *Opera genuina omnia*, appeared in Venice in 1548. The oviduct carries his name today.

30. Several ancient physicians bear the name of Soranus. The most famous was from Ephesus and practiced in Alexandria and in Rome under Trajan and Hadrian. Several medical works by Soranus were translated from Greek into Latin by Caelius Aurelianus, a compiler of the fifth century, and were included in the latter's *De morbis acutis et chronicis*, published in Venice in 1547.

31. This paragraph appears only in the 1579 edition.

32. "*Mais suiuant nottre opinion . . .* " ("But according to us") replaces *Or* in the 1579 edition to provide an appropriate transition after the insertion of the preceding Soranus interpolation.

33. To translate Joubert's Rabelaisian term *mantule* (Cotgrave gives "a mans yard. *Rab*").

34. The bracketed passage, appearing in the 1578 and in editions after 1579, was removed from the 1579 edition. The allusion is to Poggio's *Facetiae*, 62: *De Guilhelmo qui habebat priapeam supellectilem formosam.*

35. Because of the natural process of aging, as explained by the theory of humors.

36. Joubert's note: Chap. 22[: 13–21].

37. "We have a virgin."

38. Several Latin editions of Avicenna's five-book *Canon* were published, starting in 1473. On Avicenna, see note 6 of chapter II in Book One above.

Alessandro Benedetti (1460–1525) taught anatomy at Padua. He was a strong advocate of dissection and was responsible for the construction of Padua's anatomical theater. The work to which Joubert is referring is the *Anatomiae siue historiae corporis humani liber* (Venice, 1493).

Giacomo Berengario da Carpi (1465?–1550) was a physician and surgeon who wrote two works on anatomy: *Commentaria cum amplissimus additionibus super anatomiam mundini una cum textu ejusdem* (Bologna, 1521); and the *Isagogae breves perlucidae ac uberrimae in anatomian humani corporis* (Bologna, 1522).

39. Joubert first refers to Vesalius's *Anatomes totius* (Paris, 1565) and then to the *Radicis chynae usus* (Lyons, 1547). Ashes (from vine twigs) had several uses: they served as a refrigerant and as an ingredient in several remedies. See Celsus, *De medicina*, II, 33.3, and VII, 1.1.

40. This entire paragraph appears only in the 1579 edition.

41. The *arrierefosse*, a term borrowed from the vocabulary of fortification, might be translated as "last ditch"; Cotgrave gives "A backe ditch, or dike; also, as *Reffiron.*" This latter term is also clearly defined by Cotgrave: "The third gate of the wombe; or the mouth of the matrix, which is cleft acrosse, and not lengthwise, as the Hymen, &c."

42. This entire paragraph was struck from the 1579 edition.

43. The cervix.

44. This paragraph does not appear in the 1579 edition.

45. The aphorism to which Joubert is referring is as follows: "LIX. If a woman does not conceive, and you wish to know if she will conceive, cover her round with wraps and burn perfumes underneath. If the smell seems to pass through the body to the mouth and nostrils, be assured that the woman is not barren through her own physical fault."

46. The phrase beginning " . . . the practice of the Negroes . . . " and running to the end of the sentence appears only in the 1579 edition. In the original edition and in all others based upon it, the phrase reads: " . . . *à ceus d'vn païs de par le monde (il ne me souuient pas ou c'est)* . . . " translating as " . . . those [signs used] in a country somewhere in the world (I do not remember where it is) . . . "

Pietro Bembo (1470–1547), humanist and poet, was secretary to Pope Leo X. He was appointed historiographer of Venice in 1529 and wrote the work to which Joubert is referring, *Historiae venetae libri xii* (Venice, 1551); Joubert probably consulted the edition published the same year in Paris under a slightly different title (*Rerum venetarum historiae libri XII*).

47. "As laws are made, so is subterfuge."

48. Celsus, *De medicina*, VII, 25.

49. Joubert is referring to the myth of Argus, set as a guardian over Io after she was changed into a heifer. Argus was fooled by Hermes, who managed to put the monster to sleep and slay him.

50. Joubert is referring to the obscene anecdote in Rabelais's twenty-eighth chapter of *Le Tiers Livre*, in which the aging Hans Carvel, after marrying a young wife, becomes obsessively jealous. In a dream the devil, slipping a ring on the old man's finger, assures him that as long as he wears it he can be certain of his young wife's fidelity. Upon awakening from the dream, he finds his finger in his wife's vagina. Although Rabelais's source is possibly Poggio's *Facetiae* (133: "Visio Francisci Philelphi"), Barbara C. Bowen has informed me that this facetia appears in at least a dozen collections during the sixteenth century.

BOOK FIVE, CHAPTER V

1. A cupping glass (*vantouse*) was a wide-mouthed globe, usually seven to ten centimeters in diameter, in which a small amount of combustible material, such as cotton soaked in alcohol, was burned. As soon as the material was consumed, the glass was applied tightly against the skin; the ensuing vacuum drew blood to the area covered by the cupping glass.

2. This passage reveals the Renaissance idea of the *motion* of blood. Although the discovery of circulation is classically attributed to William Harvey (1578–1657), as described in his *Exercitatio anatomica de motu cordis et sanguinis in animalibus* (1628), it was Marcello Malpighi (1628–94), professor and anatomist at Pisa and Bologna and physician of Pope Innocent XII, who, in discovering the capillaries, accounted for the *circulation* of the

blood. Harvey's explanation had the blood delivered to the tissues by the arteries and gathered from them by the veins, but it did not have the vision of the *closed system* enjoyed by Malpighi. See Brabant, 229–40.

3. Cotgrave defines the verb as follows: "To swell, or increase, as womens breasts doe when the matricall veins are stretched by the menstruall bloud."

4. The derivation of *hystéra*, or *hystére*, is uncertain, but Joubert imagines a connection with *hýstera*, meaning "latter," or "last," perhaps based on Henri II Estienne, who suggests this etymology in his celebrated *Thesaurus linguae graecae* (1572). C. Alexandre's *Dictionnaire grec-français* (Paris, 1892), based on Estienne's, is the only modern dictionary to allow it. All others reject it. See Franz Passow, *Handwörterbuch der Griechischen Sprache* (Leipzig, 1841–57); Hjalmar Frisk, Griechisches Etymologisches Wörterbuch (Heidelberg, 1960–72); and Pierre Chantraine, *Dictionnaire étymologique de la langue grecque* (Paris, 1968–80).

BOOK FIVE, CHAPTER VI

1. This is an allusion to Montpellier's renown as an illustrious center of studies in medicine.

2. The second section of this sentence is highly elliptical; it should read: ". . . and the milk from a woman who had a son is better for a daughter [female child]." The obviousness and the predictability of Joubert's corrective posture must not cause us to lose sight of two crucial aspects of Renaissance thought: the relentless application of deductive reasoning and the theory of humors. Blood was related to heat (fire), black bile to cold (air), mucus to moisture (water), and yellow bile to dryness (earth). Combinations of these formed the four basic personality types: yellow bile + black bile = melancholic; black bile + mucus = phlegmatic; mucus + blood = sanguine; and blood + yellow bile = choleric, thus closing the circle of the four humors and the four elements.

3. That is, the milk coming from a woman who has just given birth to a female child.

4. Joubert's note: Objection. Solution.

5. Joubert's note: Objection. Solution.

6. Hippocrates, *Regimen* [*Peri diaites*], II, lvi.

7. Again, an elliptical sentence: ". . . hotter than the milk of a woman who carried a male child."

BOOK FIVE, CHAPTER VII

1. "Getting aroused" is regrettably not as picturesque as Joubert's expression "*s'echauffer an leur harnois*," which translates literally as "to get all heated up in one's armor."

2. This notion is reminiscent of Saint Paul, 1 Corinthians 7:9, "But if they do not contain themselves, let them marry. For it is better to marry

than to burn." Joubert goes on to argue rather convincingly against the ineffectiveness, if not the evil, of repression.

3. Joubert's note: See the exhortation in the first chapter of this book.

4. Breast-feeding while pregnant was nearly always problematic in the Hippocratic tradition. See note 6 of chapter II in Book Five above.

5. Cotgrave defines *engan[a]r* thus: "To deceiue; and (more particularly) a nurse to conceale her being with child; wherby she defrauds her charge of due nourishment. **Langued.*" The Italian *ingannare* had the same sense.

6. As in the first chapter of this book, Joubert finds himself in the uncomfortable position of arguing against authority on the basis of experience. His compromised resolution of the contradiction takes the form of postulating two types of people: the coarse but robust and healthy commoner representing what experience proves; and an educated or refined class enjoined to follow the precepts of antiquity as confirmed by the physicians.

7. The sixth book of the *Erreurs populaires* is the opening section of Part Two of the work, which appeared separately in Paris in 1579 (*chez* L'Angelier) and again in Paris in 1580 (*chez* L. Breyer). The combined parts, one and two, were published in one volume for the first time in 1587 (*chez* Claude Micard). See the Introduction.

8. *Adustion* is from the vocabulary of the theory of humors; it is derived from the Latin *adurere*, to burn.

BOOK FIVE, CHAPTER VIII

1. Cotgrave gives the following definition: "*Tandrieres*: *f.* Chaps, rifts, or chawnes on the nipple of a womans breast."

2. Cotgrave: "Pomatum, or Pomata (an oyntment) . . . "

3. This is one of the few instances in which Joubert, rather than invoking the authority of the ancients for a specific treatment, prescribes on an empirical basis.

BOOK FIVE, CHAPTER IX

1. There is an unfortunate misprint in the 1579 edition: "*aigueus*" is repeated where it should read "*aigu.*"

2. The 1579 edition reads: "*Et la frequance de l'alimant est d'autant plus requise es premiers iours, qu'il est . . .* " where the others read "*La frequance de l'alimant est requise es premiers iours, d'autant qu'il est . . .* "

3. "He who does not have a full stomach cannot sleep soundly."

4. Orlando de Lassus (1531–94), a Belgian, was among the greatest composers of his time. He lived at court in Bavaria and wrote lieder, masses, motets, and madrigals.

5. Cotgrave gives several steps for Joubert's *branle*: "A totter, swing, or swinge; a shake, shog, or shock . . . also, a brawle, or daunce, wherein many (men and women) holding by the hands sometimes in a ring, and otherwise at length, moue all together."

6. Joubert's note [1579 edition only]: *Venter famelicus non habet aures.*

The French version (*Ventre affamé n'a point d'oreilles*) appeared in Rabelais (*Le Tiers Livre*, xv).

7. Joubert's note [all editions]: *Inanis venter non audit verba libenter.* This proverb is reminiscent of Seneca: *Venter praecepta non audit* (*Epistulae ad Lucilium*, XXI, 11).

8. See note 4 of chapter VI in Book Four above.

9. Joubert's note: Objection. Response.

BOOK FIVE, CHAPTER X

1. Recalling Juvenal's line (356) in the tenth satire: *Orandum est ut sit mens sana in corpore sano.*

2. " . . . se creue [n] t . . . "; that is, develop hernias.

3. "His heart is growing."

4. According to Estienne, the Greek verb *antipelargeo* means "to pay back, to show one's gratitude": "as the stork that nourishes, as they used to say, its old parents," It is supposedly derived from *pelargos*, "stork."

5. Joubert's metaphor is " . . . *qu'elles pullule[n]t* . . . ": " . . . [lest] they put out too many shoots . . . "

BOOK FIVE, CHAPTER XI

1. Literally, " . . . requires a different type of coherency" (" . . . *requiert diuerse fasson d'antretenement*" [the 1579 edition mistakenly gives *antretemant*]).

2. That is, someone from the latter regions wanting to live in the former.

3. Joubert's note [misplaced in the 1579 edition; see note 6 below]: That is to say, she will keep growing for a longer time.

4. Joubert's note: Objection. Response.

5. Joubert's note: Objection. Response.

6. A printer's error placed at this point in the 1579 edition the marginal note that appears earlier in the text in the other editions (see note 3 above).

7. Because of a printer's error, the 1579 edition has two ampersands in this clause: "*Car l'vn & & l'autre* . . . " The printer may have wanted it to read " . . . & l'un & l'autre . . . "

NOTES TO APPENDIX A

1. This is the Table of Contents. Since it is the same as that in the Preliminary Matter section, it is not reproduced here.

2. These "Popular Sources of the First Five Books," found only in the 1579 Millanges edition, are the specific popular adages or notions upon which the discussion in the various books is based. Some of them are drawn from the Medley ["Melange"], the chaotic source of all of the *Erreurs populaires*.

3. The point of this saying is lost in translation: the word *mule* means both mule and chilblain.

4. *Nouër l'esguilhette* means, according the Cotgrave, "The charming of a mans codpiece point so, as he shall not be able to use his owne wife, or woman (though he may use any other;) Hence; *auoir l'esguillette nouée*, signifies to want erection; (This impotence is supposed to come by the force of certaine words uttered by the Charmer, while he ties a knot on the parties codpiece-point.)"

5. Cotgrave gives the following translation of this proverb: "For the one does no good, the other much hurt, if it be kept."

6. I am giving the material of Part Two as reorganized for the 1579 edition. There was to be little resemblance between it and its form in either the 1580 or 1587 edition.

7. "Every [food?] is healthful for the healthy."

8. "Quality is not harmful, but quantity is."

9. "After something bloody [or raw], someting undefiled [pure, clean]."

10. The rhyme of "*salade . . . malade*" is lost in translation.

11. The rhymes "*pomme . . . homme*" and "*poire . . . boire*" are lost in translation.

12. "*Fourmage . . . mariage*": lost in translation.

13. "*Lait . . . souhait*" and "*vin . . . venin*": likewise lost.

14. *Faire la grasse matinée* is still the common French idiom meaning "to sleep late into the morning."

15. See chapter IV of Book Four.

16. See the Popular Sources of the First Five Books above (Of the First, No. 12).

17. *Potus* is a general term for liquid remedies of which the basic solution was wine. I am not certain about the meaning of *semen contra*; it is possible that it refers to remedies based upon seeds of plants having virtues contrary to the perceived nature of the illness being treated.

18. Cotgrave defines the "Climatericall" years as follows: " . . . euerie seuenth, or ninth, or the 63 yeare of a mans life; all very dangerous, but the last, most."

19. Possibly an Occitan saying meaning "He who is late in arriving is late in taking leave."

NOTES TO APPENDIX B

1. Etienne Maniald was a Hellenist who participated in the publication of the Greek texts of Hermes Trismegistus. Millanges published Maniald's *Mercurii Trismegisti pimandras* in 1574.

2. Jean Fernel (1496–1558) enjoyed the reputation of a Renaissance man: in his *Cosmotheoria* (1528) he measured with considerable accuracy a portion of the meridian; he wrote several medical works, among which one often cites his work on syphilis, *De luis venereae curatione perfectissima liber* (Antwerp, 1579); and he is credited with curing the sterility of Catherine

de' Medici when he was the personal physician of Henry II. See note 19 in chapter IV of Book Five.

3. *Epiones* is possibly a noun based on the Greek adjective *epios*, which means "gentle" or "mild" and which also is applied to remedies that are soothing or assuaging. On Sylvius, see note 19 in chapter IV of Book Five.

4. Dominique Revlin, a physician from Bordeaux, as indicated in the title of his epigram, published a book on surgery in Paris in 1579 (*La chirurgie de Dominique Reulin*). He also published in the following year in Montauban his *Contredicts aux erreurs populaires de L. Joubert*.

5. This is possibly René Guillon (1500–1570), who wrote a Latin translation of the letters of Isocrates (*Isocratis oratoris Atheniensis, epistolae graecae* [Paris, 1547]). He also published works on grammar and prosody and wrote the *De dialectis verborum et nominum* (Paris, 1561) and the *Tabulae monstrantes viam qua itur in Graeciam* (Paris, 1567).

6. Pierre de Brach (1548–1605) was a poet and lawyer from Bordeaux whose principal works, *L'Aminte* (1584) and *La Jerusalem* (1596), have not survived the test of time. See Sabatier, 233–34.

7. Nothing is known of Joseph Du Chesne other than the information furnished in the *Erreurs populaires*: he was a doctor of medicine and proprietor of a domain called Liserable, and, if his liminal poem reflects the truth, as Joubert's former tutor he seems to have had a paternal affection for the author.

8. Pierre Chambon de Gotz does not appear in any of the standard bibliographies or biographies, nor have I been able to find mention of him in the edited correspondence of the time that was at my disposal.

9. Salomon Certon (1550–1610) was a French poet and translator who put both the *Odyssey* (Paris, 1604) and the *Iliad* (Paris, 1615) into French. See Sabatier, 234–35.

10. This is a translation of *los me coronant*, an anagram of Salomon Certon. Sixteenth-century writers frequently made anagrams of their names and signed them to their works, either out of fear, as in the case of François Rabelais (Alcofribas Nasier), or in an attempt at cleverness, as is the case with Certon in the present example.

11. Simon Millanges was the famous Bordeaux printer who published, among many other illustrious works of the time, Montaigne's *Essais*. See the Introduction.

NOTES TO APPENDIX C

1. In the 1579 edition Millanges replaces the remainder of the letter with the following sentence: "They will also excuse me, should it please them, if, unaccustomed to the orthography that Monsieur Joubert wants respected in his books, I allowed several words to be printed in different ways and otherwise than the aforementioned Sire Joubert wished."

2. Millanges was henceforth to print only the 1584 edition, which is technically nothing more than a reprinting of the 1578 edition.

3. The only chapter Millanges marked with an asterisk was chapter IV of Book Five.

4. See the Table of Contents in the Preliminary Matter and also Appendix A.

NOTES TO APPENDIX D

1. Dion Cassius was a Roman historian of the second century A.D. who wrote several historical works of which the most famous was a history of Rome (*Romaike historia*) in eighty books, but of which only portions are extant, surviving in the form of Xiphilinus's epitomes. Joubert gives Book 58 as reference in a marginal note.

INDEX OF NAMES, PLACES, WORKS, AND SUBJECTS, INCLUDING ALL ANATOMICAL, MEDICINAL, AND PATHOLOGICAL TERMS